Warp Knitted Fabrics
Construction

Warp Knitted Fabrics Construction

Yordan Kyosev

ITM – Institute of Textile Machinery and High Performance
Material Technology
Chair of Assembling Technology for Textile Products
Technische Universität Dresden
Dresden, Germany

CRC Press
Taylor & Francis Group
Boca Raton London New York

CRC Press is an imprint of the
Taylor & Francis Group, an **informa** business

CRC Press
Taylor & Francis Group
6000 Broken Sound Parkway NW, Suite 300
Boca Raton, FL 33487-2742

© 2020 by Taylor & Francis Group, LLC
CRC Press is an imprint of Taylor & Francis Group, an Informa business

No claim to original U.S. Government works

Printed on acid-free paper

International Standard Book Number-13: 978-1-4987-8016-2 (Hardback)

Visit the Taylor & Francis Web site at
http://www.taylorandfrancis.com

and the CRC Press Web site at
http://www.crcpress.com

Warp knitted fabrics are a high achievement of the textile engineering spirit. Their variety, complexity and beauty remain until now deeply appreciated only by a small number of leading experts in this area. This book is dedicated to the current and future warp knitters, with the wish of helping them become experts and to help keeping and extending knitting knowledge.

Contents

Foreword

Warp knitted fabrics are a very important class of textile products which are manufactured with highly efficient and productive warp knitting technology. Although most of warp knitted fabrics are produced with continuous filament yarns, their end uses are extremely wide and have rapidly been expanded from traditional apparel, home textiles to highly technically required products. Lingerie, lightwear, sportswear, bedsheets, dressing gowns, outerwear, shoes, domestic and automotive upholstery, curtains, drapes, laces, geotextiles and composite reinforcements are examples of end uses. This large variety of warp knitted products has made the design of the fabric structures a very complex task and knowledge about warp knitted fabric construction has become a necessity for the people working in this area.

To my knowledge, the book *Warp Knitted Fabric Construction* written by Prof. Yordan Kyosev is the first English book particularly focusing on warp knitted fabric structures. The book covers all aspects of warp knitted fabrics from the fundamentals of warp knitting to the use of software to design warp knitted structures. Although there have been various books on warp knitting, these books are old and most of them are not available for many of new readers. The new book provides a number of simulated images of fabric structures created from *Warp Knitting Pattern Editor 3D* based on the author's many years of research on 3D modelling of warp knitted structures. The free available TexMind Textile 3D Viewer software for the better visualisation cannot only help readers easier understand warp knitted fabric structures, but also increase their interest in warp knitting. In addition to being used as a textbook for university students, the book can be a good reference book for engineers, designers and other relevant people working in this area.

Prof. Hong Hu
Institute of Textiles and Clothing
The Hong Kong Polytechnic University

Preface

Warp knitting is a highly efficient and productive fabrics production process. Due to the large number of possible combinations between the lapping motion, threading and guide bar arrangement, the design of the structures remains a very complex task.

The first deep contact to warp knitting after the regular engineering courses at the Technical University of Sofia, was in year 2000. In this time I was a visiting scholar at the Niederrhein University of Applied Sciences, Mönchengladbach, Germany and was able to participate in the latest warp knitting lessons of Prof. Dr. h.c. Klaus-Peter Weber. I was sitting as a guest student, but making notations as a lecturer, as I had to teach this content later in my home university. His books helped me understand the fundamentals of the process and the warp knitting pattern, and I have taken his enthusiasm for warp knitting and tried to transfer it to my students. From 2001 to 2004 I had to lead the practical exercises in this area and in the academic year 2004/2005 the lecture courses in Technical University of Sofia.

During my active work in the 3D modelling of textile structures I met in 2006 with Dipl.-Ing.(TH) Wilfried Renkens, the person who I have to thank for the very good collaboration - he caused me to go into deep details of the 3D modelling of warp knitted structures.

Working for and with Mr. Renkens I developed in 2006-2007 the first engine for 3D visualisation of warp knitted fabrics, demonstrated by the company ALC Computertechnik (Aachen) (now Texion) on the ITMA Munich 2007. Later the ALC Computertechnik was dismantled. There were different requests for improved or new 3D solutions in the area of warp knitting. The first version of Loop3D and its integration with the new Warp Knitting Pattern Painter was born. Today both of these are history, too and are replaced by a next generation of package Warp Knitting Pattern Editor 3D, developed and maintained by TexMind UG and used for the creation of all *simulated* images in this book.

The goal of this book is to collect at one place the available knowledge in the area of warp knitted fabrics and to make this knowledge more understandable using the modern tools for visualisation. This knowledge exists spread out in several older books, which are not available or within the brains of elder people. In short extended knowledge about warp knitted structures is not ***publicly available*** in a well structured way.

Except for the book of Professors Weber and Weber [26] , where the processes are explained with a lot of photos and paintings, the remaining books are available only through an antiquated market. The good old fashioned warp knitting book with real samples like this from Weber [25], from Rogler and Humboldt [23] gives real impression about the fabrics, but probably will never be produced in this form. As partial replacement here an idealized 3D simulations of some fabrics are attached, so that the reader can rotate, zoom and better understand the yarn positions there. Great books about warp knitting are those of Paling [18] and S. Raz [19]. I like as well very much the books of my colleague Arsu Marmarly from Turkey [14] and the team of Prof. Kopias [5] from Poland, written in the current time, but these are available probably only to the their local students or if you contact them directly, and if you can read the corresponding language.

I have learned a lot from the books of Mr. Wilkens[28] who unfortunately passed away this year. The combination of his extended knowledge and experience, and the willingness to write down it in a book and my software and computer skills would be the best for a team for a warp knitting book, but though it was intended, we never meet personally.

I like to thank all persons who helped me in any way in the development of better understanding of the warp knitting structures:

- One of those persons is Mrs. Beate Pfeifer, ***the warp knitting expert***, who I met several times personally, for the nice talks about the drive and the motivation to go further and further in 3D modelling. Actually, as a real "full time job" expert, she never spoke about the structures with me and but I as well never offered her to pay her for such conversation.

- Mr. Meyer, who worked until and probably after his retirement, too for the company Müller Textil with his great knowledge about the spacer fabrics and who motivated me to think about new methods and models for the 3D simulation of warp knitted structures.

- I like to thank my former students, Ms. Eugenia Bomm, Mr. Matthias Aurich, Ms. Ruth Neumeier for their intensive testing of the different versions of the software and supported it with user's guides.

- My special thanks is to my assistant Mr. Frank Heimlich, who was always able during the lunch breaks to discuss with me complicated questions, about the positions of yarns in the loops and underlaps, the machine part orientations or historical issues in the numbering, etc.

- I would like to thank as well the editorial manager of Taylor & Francis Books - Dr. Gagandeep Singh - who had the patience systematically to bother me, in its best positive meaning of this word, until I finally agreed to find time in my very busy schedule, then to start this book project and at the end to finalize it.

My intention at the beginning was to write only 1/4 of this book - but the initial other three co-authors left me alone in this project, because their professional obligations or healthy conditions did not allow them to contribute. As time management does not allow more time for this project now, you have in your hands the **first** edition of one work with all its advantages - including new images, 3D views, as well with teething troubles. Due to my limited time, it was not possible for me to simulate all structures with suitable following relaxation, so the 3D images presents **idealized** structures. I had to skip several structures and large number of interesting samples. Nevertheless, I hope that the book in its current form will help enough people to fall in love with warp knitting or at least simplify their teaching or learning process.

For better visualisation, part of the structures is included as files in the appendix. These files are in the TexMind native XML format, so that the readers can open, rotate and zoom these images with the free available TexMind Textile 3D Viewer software.

I would be really thankful for any remarks and comments about the content, which can help to get extended and probably a better edition in the near future.

January 2019, Mönchengladbach, Germany
Yordan Kyosev

Author

Yordan Kyosev holds the M.Sc. degree in "Technique and Technology of the Textile and Clothing" (1996), the M.Sc. degree in "Applied Mathematics" (2002), and a PhD in Textile machines (2002) (Technical University Sofia, Bulgaria). He attained habilitation in the area of "Textile Technology" in Technical University Chemnitz, Germany in 2018. Between 1996 and 2005 he was an assistant professor for Textile Technology at the Technical University Sofia, Bulgaria (design of textile and sewing machines, knitting technology and technical mechanics). After a research stay with Post-Doctoral Fellowship of the Alexander von Humboldt foundation at the Institut für Textiltechnik of RWTH Aachen University in 2005-2006, he became a Professor of "Textile materials, textile technology and quality management" (2006) at the Faculty of Textile and Clothing Technology, Niederrhein University of Applied Sciences, Germany. During 2019 he joined Technische Universität Dresden and took over the chair for assembling technology for textile materials. He developed the 3D module of the first industrial software for 3D visualisation of warp knitted fabrics for ALC Computertechnik (2006-2008). In 2011 he founded the software company TexMind (www.texmind.com), specialized in the development of algorithms for modelling of textile structures, software "Braider", "Braiding Machine Configurator" and "Warp Knitting Pattern Editor 3D", used for creating the most figures in the current book. Prof. Kyosev is an author of numerous publications on modelling of textile structures, braiding and knitting technology, and fuzzy logic. Since 2018 he is an Editor-in-Chief of the "*Journal of Engineered Fibers and Fabrics*".

Symbols

Symbol Description

GB	Guide bar	CPC	Courses per centimeter
GBi	Guide bar with number i	CPI	Courses per inch
JB	Jacquard guide bar	WPC	Wales per centimeter
PB	Patterning guide bar	E	Machine gauge
FA	Distance between the needle beds on double needle bed machine		

Part I

Fundamentals of warp knitting

1

Warp knitting process fundamentals

1.1 Introduction

This chapter provides information about the warp knitting process, limited to the *patterning* related topics. More details about the knitting process, the working cycle of the machine, different configurations of machine elements and so on can be found in several books such as [4], [24], [26], [18].

1.2 Knitting process and pattern notation

Each warp yarn, used for production of warp knitted fabrics, is formed from a single guide. Several guides are grouped into guide bar, as shown in Figure 1.1. Each guide bar moves around the needles and places the yarn over, between and after or behind these during the production process. The needles, the spaces between them and the guides are numbered, starting from the side, where their pattern drive (cam, chain links, electronic drive) is mounted. Usually, the pattern drive is placed on the right hand side of the machine and the needles, and the guides are numbered from right to the left. For the graphical notation of the warp knitted structures, the complete motion process of the guides is recorded symbolically on paper. For this notation, the person has to assume, that is looking at the needles from the top (Figure 1.1 upper picture), but for simplification, only the needle cross section is drawn as a thick point. For each knitting cycle, one row with points and spaces between these is used.

Let the guide bar have only one yarn on its guide number one (Figure 1.2). During the knitting process, the guide moves **between** the needles and this motion step is named "swing in"-the needles motion. The gab number, where the **first** guide of one bar is swinging-in, is used for numerical notation of this motion step; in the case of Figure 1.2, this is the number 3. On the drawing, a short line (or curve) between the dots is drawn.

After the placing the yarn(s) between the needles the guide bar is moving to left or right in the front of the needles, placing yarn over them - overlapping (Figure 1.3). The overlapping is a shog over one or seldom over two needles. The limitation to one or two needles is based on the need of the yarn length

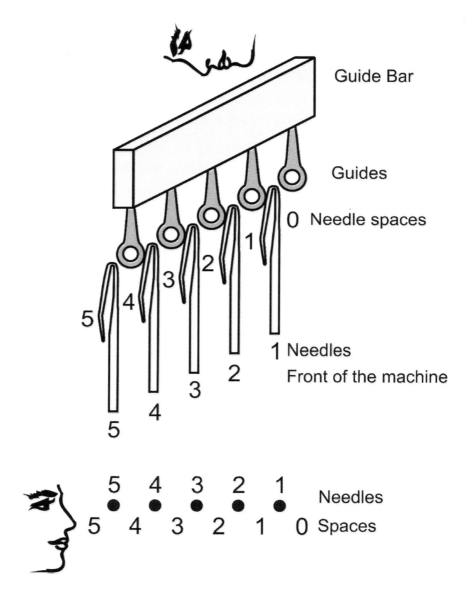

FIGURE 1.1 Notation principle of the warp knitting pattern.

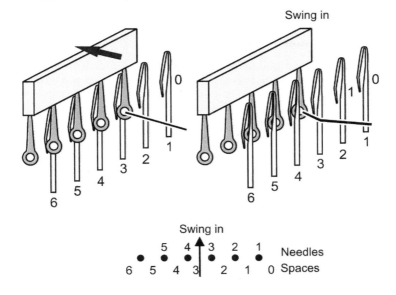

FIGURE 1.2 Swing in motion step.

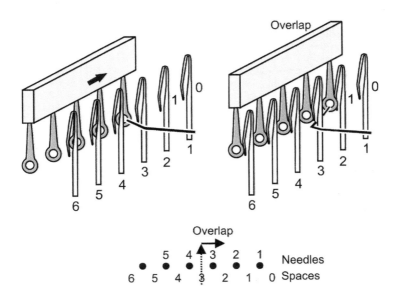

FIGURE 1.3 Overlap - the motion of the guide bar placing yarns on the beard side of the needles.

for the loop building process. If yarn is placed over more than two needles, during building of the loops there will be not enough yarn length for more loop heads, legs and underlaps and the yarn will break. In pattern drawing this step is drawn as a small arrow or arc over the point. If no loop has to be built, then the guide bar stays there and does not perform any motion during this step.

The next step is the swing out - the guides are swinging through the gaps between the needles back. The gap number, where the first guide goes through is used for the numerical notation of the pattern, in the case of the Figure 1.4 this is the number two. On the drawing, this motion is represented by a short line or curve between the points.

Finally, the guide bar moves behind the needles and the guides place their yarns "under" the needles (Figure 1.5). The motion is named underlap. This motion is not limited to one or two needles, because the yarn does not have to build loops on this side, and the required yarn length has to be taken from the beam. The limitation of the underlap motion is determined from the patterning device - it can be often not more than 32 gaps.

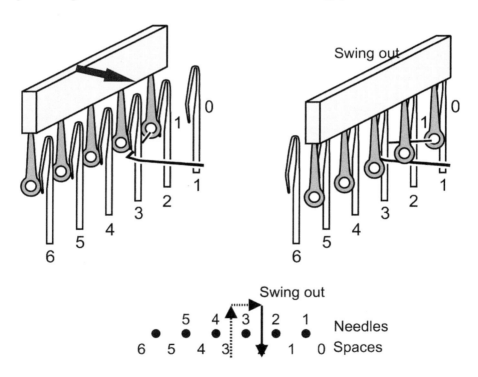

FIGURE 1.4 Swing out - move the yarns back between the needles.

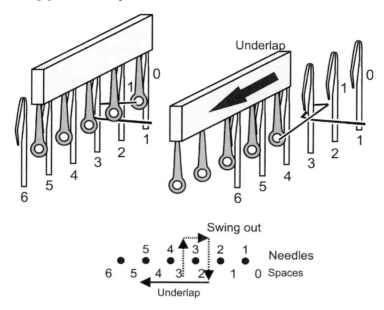

FIGURE 1.5 Underlap - place the yarns behind the needles.

After these four steps of the guide bars, the needle bar starts moving all the needles down. Theirs hook have to become closed. Depending on the needle type this happens as follow:

- the bearded needles become pressed during this period;

- the latch of the latch needle rotates, moved from the previous loop;

- the tongue of the compound needle moves and closes the hook.

The new placed yarn piece in this case remains in the needle hook and the previous loop can knock over it. After that, the needle bar moves up and the cycles start again. For each such cycle, the numbers of the swing-in and swing-out positions are notated, separated by minus character like "3-2", the numbers from the next cycle are separated by a slash "/". In this way, the motion, depicted in Figure 1.6 is coded as "3-2/5-6/3-2/0-1//". The end of one repeat is marked with two slash lines "//". For each cycle a new row with dots for the needles is drawn. The swinging and under- and overlap motions are normally not running as straight lines; there is smooth transition between these as both motional directions are performed from independent devices. It is as well more convenient for the painting manually to draw the path smooth, so the symbolically represented motion of the guide of Figure 1.6 a is represented normally as the smooth red line of Figure 1.6 c.

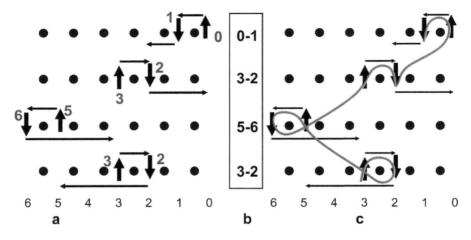

FIGURE 1.6 Graphical representation of the lapping movement.

1.3   Open and closed loops

The way, in which the curve goes around one needle during each cycle, already contains information about the type of the loop in the structure, as represented in Figure 1.7. On the left hand side of this figure, the previous motion is given again, but in the way as it is drawn (without the explanation arrows for the motion steps). The needles respectively the points, around which the curve is crossed, are building closed loops; these are in the case of the first and second cycle (3-2 and 5-6). If the curve, drawn by the guide, does not cross itself around one needle, the loop is named open loop. Such are the last two loops (3-2 and 0-1). The type of the loop depends not only on the swing-in and -out positions at the current cycle, but as well on the initial position of the guide during the next step. If the after the overlapping motion, the guide continues the next underlapping motion in the *same* direction, then an *open* loop is built, because the legs of the loop do not cross, these are pulled apart. If after the overlap motion the guide changes the direction for the following underlap, then a *closed* loop is built, as the loop legs cross (Figure 1.8) and are moved in together.

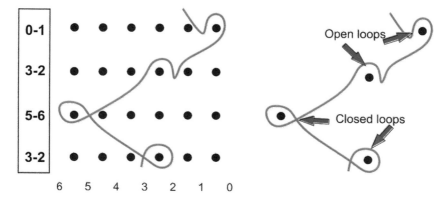

FIGURE 1.7 Graphical representation of open and closed loops.

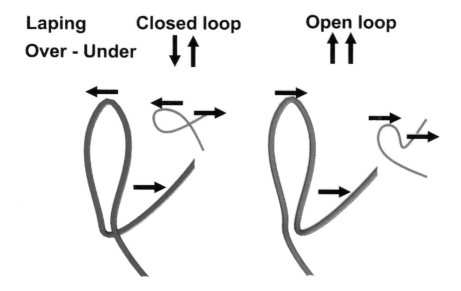

FIGURE 1.8 Open and closed loops - if both the overlapping and the following underlapping are in the same direction, the loop is open (right); if these are in the opposite direction the loop legs cross and make the loop closed (left).

1.4   Numerical notation issue

Usually the numerical notation of the lapping motion was written manually on sheet of paper from up to down, with one number per row, and after each cycle the horizontal line is drawn (Figure 1.9 a). This is really confusing and not efficient, when larger laps have to be followed, because the lapping motion is drawn graphically from down to up (Figure 1.9 c), which means in the opposite direction. In the modern computer age, there is no problem in writing down the numbers for each course at the same position, where the course is, so this system will be used in this book (Figure 1.9 b). In this case, both the graphical and numerical notations are running synchronously from down to up. Longer laps are noted in a text row, under the picture.

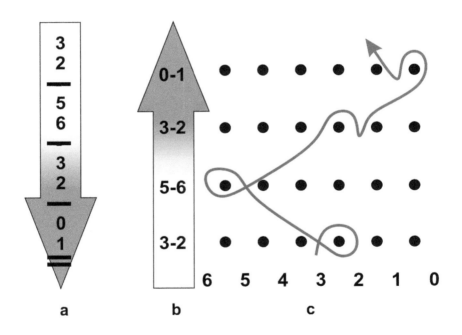

FIGURE 1.9 Chain notations: a) commonly used, but not a convenient system - numbers from up to down, pattern down to up; b) used in this book system - the swing-in and swing-out motions are given in the same line as the corresponding loop.

1.5   Connecting loops in spacer structures

There is a third type of loops for machines with two needle bars. The **connecting** (pile) yarns there build loops on **both** needle bars and their legs are going almost perpendicular to the loop plane. In this situation the legs do not cross for both - open and closed loops, they go to the next needle bar. These loops can be named **connecting loops**, see Figure 1.10. Some particularities about the loop types for double needle bar machines are discussed in Chapter 10.

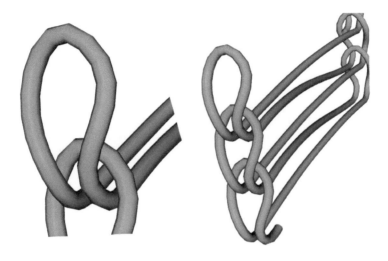

FIGURE 1.10 Connecting loop of a double needle bar machine, as a third type of loops.

1.6   Tempi- 2- and 3-, 4- and 6- tempi lapping movements

Three tempi are lapping movements for machines, where the motion of the guide bars is arranged into three steps for one loop building cycle. Such machines were used in the past, usually with mechanical chain links for control of the guide motion with the purpose of creating a smoother motion law of the guides. The first two steps of the motion are remain with the same goal as of the two tempi machines- the guides swing in between the needles and swing out at the same or other position. The additional third step is used to split the underlapping into two sub-steps and in this matter to allow more time for longer underlaps. For the structure the third number does not have any influence, it has influence on the smoothness of the guide motion.

The two tempi lapping in this way 1-2/3-2// could be written as 1-2-2/3-2-2//, which in reality does not give advantage for the machine, because the guide will stay at the same position, but it can be written as well as 1-2-3/3-2-1// where during the third step, the guide will move to the position for the next one and will have better dynamics.

More sensible is the meaning for pattern with larger underlaps like 1-2/5-6// which in three tempi machine can be written to 1-2-4/5-6-3// so that the speed of the guide bar when moving behind the needles is reduced and complete machine dynamics are better.

With the same principle for machines for double needle bars - the normal four-tempi lapping becomes six-tempi, because one complete cycle includes producing of loops of the front needle bar (first three tempi) and rear needle bar (the second three tempi).

The older double needle bar machines use only even numbers for counting the needle gaps so that the lapping 1-2/1-0// will be noted 2-4/2-0//. In this book, the gap number is always used, as defined in the current norms. For the owners and users of older machines, the software TexMind Pattern Editor provides automatic conversion from one system to the other and vice versa.

1.7   Threading

For the representation of the guide bar motion the path of its first guide is drawn, independent of whether this guide has yarn or not. The *threading* - is an information set about the type of the yarns and their arrangement in the guide bar. If the guide bar has full threading, all its guides have yarns, as shown in Figure 1.11. The needles remain on their positions, but the yarns are placed between the needles starting from the gap number three in this case.

If only some of the guides have yarns or different yarns are used, the complete coding sequence has to be noted and provided with the documentation of this sample. Each yarn has in this case a character code from the Latin alphabet and the empty guides are marked with a dot "." (Figure 1.12). The sequence "AB.AB" can be then used for building the pattern drawing - the curve for each different character is placed with different colour and the gaps, to which a dot is corresponding, do not receive any curve.

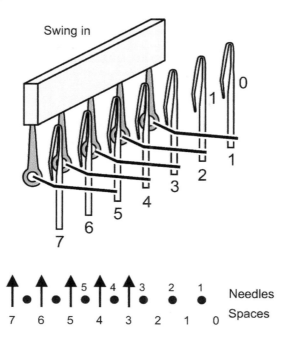

FIGURE 1.11 Full threading of guide bar.

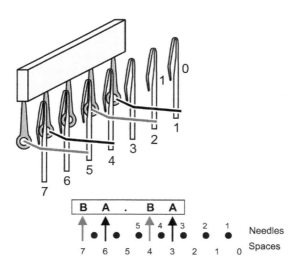

FIGURE 1.12 Partial threading with two types of yarns.

1.8   Left and right hand side pattern drive

Depending on the position of the patterning device, the numbering of the needles, needle gaps and guides can start from the left or right hand side. Most former machines of the company LIBA had the patterning device on the left hand side of the machine, while these of Karl Mayer are on the right hand side([28], p.73). For drawing on the pattern, the initial position (0) is corresponding on the right hand side (Figure 1.13 c - left) or the left hand side ((Figure 1.13 c - right)). Identical is the notation for the threading (Figure 1.13 b). Attention has to be taken during the preparation of warp beams, if both types of machines are available in the same company - as the threading of the yarns for machines with left and right placed pattern device (Figure 1.13 b), because in this case the guide numbers change their direction, but the yarns have to remain in the same arrangement.

Documentation for the threading operator as well has to consider the place on the machine, on which he/she is working. If the operator enters the yarns into the guides from the back side of the machine, he/she has to receive a mirror image of the threading plan.

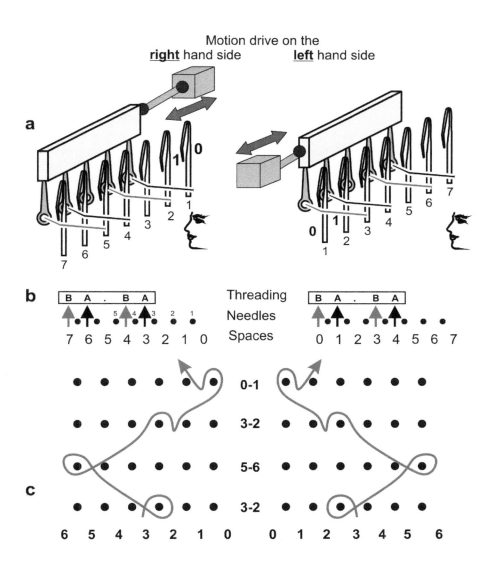

FIGURE 1.13 Right and left placement of the pattern drive.

1.9   Guide bar numbering

Guide bar numbering was different on the different types of machines during previous years. The numbering of the Raschel machines and of the Tricot machines was in the opposite order, which made the reading of older books confusing, especially if the front and back guide bars are mentioned.

The current valid norm DIN ISO 10223:2005 unified the numbering of all machines. According this norm, both Tricot (Figure 1.14 a) machines and the Raschel (Figure 1.14 b) machines start the numbering of the guide bars from the farthest in the front (or bottom) staying bar (Figure 1.14). Additionally, the type of the guide bar is mentioned, as in the general case the guide bar has the character B and a number as B1, B2. The bars, which produce loops (ground) are named Ground-Bares with abbreviation GB. The Jacquard bars, where each guide is individually controlled, are abbreviated as JB. Bars, used for effect (schuss) patterning and have the abbreviation PB and those for inlay yarns, FB.

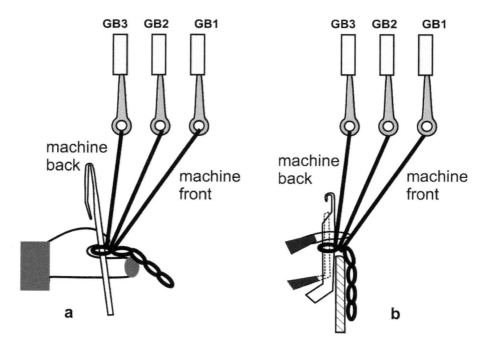

FIGURE 1.14 Unified guide bar numbering according to DIN ISO 10223:2005 a)Tricot machine, b)Raschel machine.

1.10   Fineness

The number of the needles on the needle bar in one inch (25.4mm) determines the gauge E of the warp knitting machine. The gauge is the main machine parameter related to the fineness of the produced fabrics.

$$E = \frac{Number\ of\ needles}{25.4mm} \tag{1.1}$$

aS the gauge determines the number of the needles and the needles and the sinker platines have some minimal required thickness, it determines directly as well the thickness or the fineness of the yarns, which can be used on this machine.

1.11   Courses, wales, rack, run-in

Each single needle of the needle bar(s) can build loops, if they receive yarns and a previous loop is available. All loops of one and the same needle build a vertical column named **wale** -(Figure 1.15). The loops of all needles, built during **one knitting cycle**, build a horizontal row of loops, named **course** (Figure 1.15.

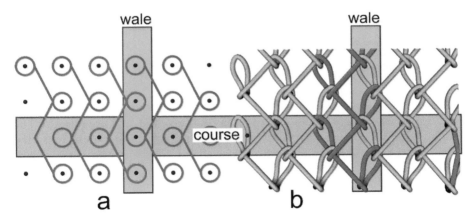

FIGURE 1.15 Courses and wales; a) on lapping diagram, b) simulated structure.

A sequence of 480 courses build one **rack**. The rack is used normally as basis for the yarn length calculation. The yarn length per single guide, required for the production of one unit length of the fabric is named **yarn run-in** and is defined usually in millimetre per rack. The run-in influences the stability of the knitting process and the appearance of the fabrics significantly, because it determines the yarn tension during the knitting and the following relaxation processes directly. The number of the courses(loop rows) per inch or centimetre (courses density) is determined by the **take-off** speed, which determines the length of the ready fabrics taken off and out of the knitting area. If the unit length is in inches, the character I is added at the end of the variable name (Courses Per Inch - CPI), and if the unit is in centimetres - the character C (Course Per Centimetre - CPC). The density of the wales (loops per unit length is equal to wales per inch (WPI) or wales per centimetre (CPI)) on the machine and is directly connected to the distance between the needles, which is related to the fineness of the machine. After the knitting process the fabric relaxes and passes the thermofixation process, which determines its final density of wales and courses per unit length.

The stitch density, SD, describes the total number of loops in a measured area of fabric. For centimetre scale

$$SDC = WPC \cdot CPC \tag{1.2}$$

and for the inch scale

$$SDI = WPI \cdot CPI \tag{1.3}$$

1.12   Face and back sides of fabric and loop

During the knitting process, the faces of the loops remain hidden for the knitter, as these remain oriented to the machine (Figure 1.16). The operator sees the back sides of the loops with the overlaps, and this because this side shows more differences depending on the pattern; it is named **technical face**. The loop side, where the loop faces are visible builds then the fabric's back. This is very confusing for new knitters, but it is easy to remember - that the loop face and the fabric face are in opposite positions.

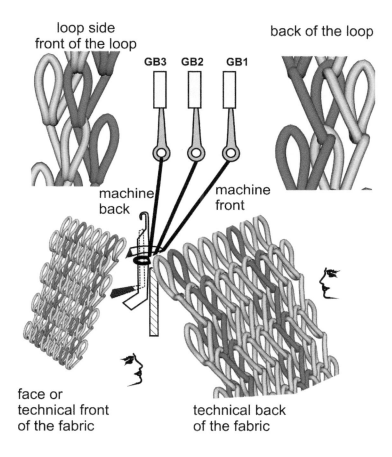

FIGURE 1.16 The face of the loop builds the fabric's back, the back of the loops builds the technical face of the fabric.

1.13   Single and double face fabrics

Fabrics, which consist of one face side with underlaps and one back side with visible face loops are named **single face** fabrics (Figure 1.17). These are produced normally on a machines with single needle bar. Of course their production is possible on machines with two needle bars, too, but it is inefficient, because half of the cycles of the guide bar motion in swinging between the needles from the second needle bar (without overlapping!) are not used for the production. Production of single face fabrics on machines with double needle bars is done only if some additional special effects are desired, where the needles from the second bed are required - as for instance for the production of hollow structures, where each needle bar produces its own single faced product. If the both fabrics - from the two needle bar machines become connected by some yarns, the fabrics become **double face** fabrics (Figure 1.17). Well-known such examples are the **spacer fabrics**, but there are as well a large number of fabrics such as meshes, heavier fabrics and other examples, which are produced as double face structures.

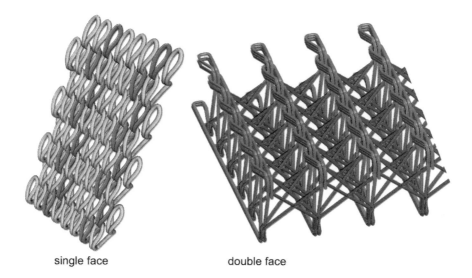

single face   double face

FIGURE 1.17 Fabrics from one needle bar have one side with loop backs (underlaps) - single face; (most, but not all) fabrics from double needle bar machines have two faces and are double face fabrics.

1.14   Repeats

The smallest number of lapping movements of one guide, which repeats in the fabrics is the vertical repeat (Figure 1.18). For the smallest atlas lapping (Figure 1.18 a) it is four courses, the tricot-stitches (Figure 1.18 b) - two courses and the open and closed lock-stitches have a repeat of one course (Figure 1.18 c). The number of needle distances, which one guide passes over during the lapping - or the number of the wales, in which its yarn participates, builds the lapping width. The atlas lapping in (Figure 1.18 a) has a width of three wales, the tricot - two wales and the lock-stich - one course.

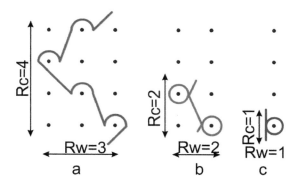

FIGURE 1.18 Vertical repeat and lapping width for a) smallest atlas, b) tricot, c) closed lock stitch.

The repeat of the threading is as well important, especially if the different guides have partial threading or threading with different colours or materials.

The total repeat in both vertical and horizontal directions for machines with multiple guide bars is calculated as the **lowest common multiple** (LCM), of the repeats of the single guide bars. The smallest common multiple of two or more numbers (integers) is the smallest positive (integer) that is divisible by all these numbers. Figure 1.19 demonstrates this for the vertical direction. The sample repeat $R_{c,sample}$ is in this case equal to the repeat of the longer lapping of the guide GB2 $R_C = 8$, because it is as well the lowest common multiple of the repeats of the both lappings:

$$R_{c,sample} = LCM(R_{c,i}) = LCM(2,8) = 8 \qquad (1.4)$$

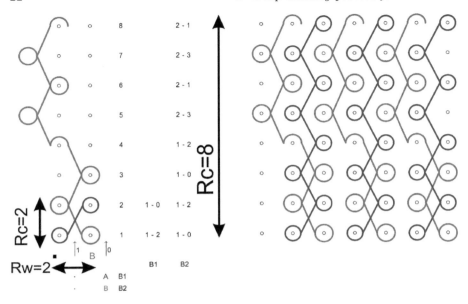

FIGURE 1.19 Repeats of two guide bar samples without full threading.

The repeats in the horizontal direction is in this case $R_W = 2$ because both guide bars have the same threading

$$R_{w,sample} = LCM(R_{w,i}) \tag{1.5}$$

1.15   Conclusions

This chapter provides the fundamentals of the warp knitting process, related to the structure of the fabrics. It does not cover in detail the stitch building and the machine elements, as these are not the object of the current book. Information about these topics can be found in the book of Professors Weber and Weber in German language in [26], the book of Professor Offermann and Mr. Tausch-Marton [17] by Springer, in Turkish in the book of Prof. Arzu Marmarali [14], in Polish and English in the book of the team of Lodz University [5]. All these great books are either not easily accessible because of the local marketing of the publishers and/or readable only for people who know the corresponding language. For English-speaking complete beginners in knitting the book of Spencer [24] can be recommended.

2

Lapping movements and stitch types

2.1 Introduction

The lapping movement of the guide bars determines the type of interlacement of the yarns, and in this way most of the properties of warp knitted structures. Additionally to the lapping movement, the position of the guide bar on the machine and the threading has as well an important influence on the structure, but their influences and combinations are subjects of the following chapters. This chapter concentrates on the lapping motions only and explains the main types of motion movements of the guide bars. In the case where the lapping movement leads to a loop building, the name of the lapping motion is the same as the name of the "stitches". For instance, closed tricot lapping means that the guide bar produces closed tricot stitch.

In the following chapter the lapping motions are divided depending on the their **structural** tempi. The tempi of one machine is not always equal to the tempi of the lapping motion. On a machine with one tempi, the guide bar makes only one motion step per loop building cycle. Several older machines are three tempi, so the guide bars have three motion steps per knitting cycle, but most machines today are two tempi.

Related to the building of the different **structural elements of the fabrics** , not all two and three tempi lapping motions are used for yarn placement, some of them are making mislapping (zero lap) not laying yarn over the needles. For instance the lap "1-1//" on two tempi machine and "1-1-1//" on three tempi machine produces the same effect as "1//" on one tempi machine. For this reason, in this chapter such laps are classified as "one tempi" lap. The 1-Tempi laps are common for the insertion of weft yarns, mainly on crochet knitting machines, but as well on some modern computer controlled machines.

The two tempi laps are the most commonly used laps, applied for all kinds of loop-based structures on single needle bed machines, and four tempi are the laps for the double needle bed machines. Some of the older machines use three and six tempi laps, explained here to provide the information for people working as well on older machines and for the sake of completeness. Special 5-6-7 tempi are used in special devices for decoration.

2.2 One tempi laps for laid-in or weft insertion

In several cases some yarns do not have to build loops during the knitting production, and the yarn guides swing-in and swing-out through the same gap between two needles. In crochet knitting machines there are several guide bars, which are performing only such motion and their programming or cams require **only** the gap number, written once. As there is only **one** horizontal motion of the guide bar for the knitting cycle, these movements belongs to the 1-Tempi group. In the case of 2-Tempi machines, both positions have to be specified, but these numbers are the same.

Most simple lapping movement is just "no movement" or "no lapping" (Figure 2.1 a) in which the guide bar stays in the same position. This motion is used to leave vertical yarns, for instance, to limit the elongation of the structure. Such vertical yarns cannot stay connected with the structure alone; they have to be integrated between other structural elements as underlaps or other wefts.

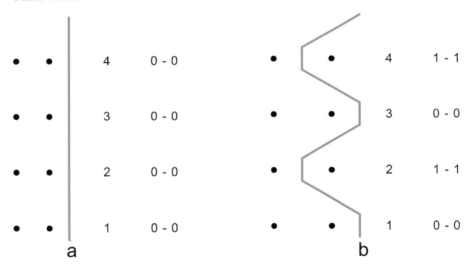

FIGURE 2.1 a) Lapping movement 0-0// ; b) Lapping around one needle, usually applied to connect elastic yarns around the loops or just to hide a yarn between the regions 0-0/1-1//.

If the guide bar moves between two positions, the yarn remains connected around the loops, produced by one needle (Figure 2.1 b). Such motion is used often to connect hyperelastic yarns (rubber, elasthan) to the remaining loops or in the transition between two areas in order to have less visible, but well-connected yarn.

For the connection of a larger number of wales (vertical column of stitches), the guide can go alternatively to change its position between two far away positioned needle gaps (Figure 2.2). In some cases, the guide bars can have a simpler pattern device, using excentric drives instead of chain links. Such are for instance often the last two bars of crochet knitting machines.

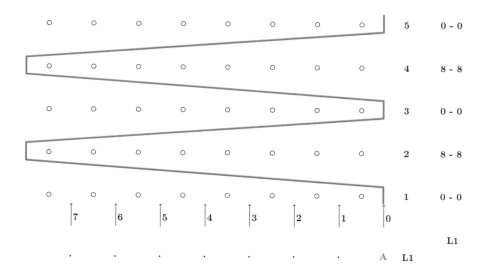

FIGURE 2.2 Lapping through a large number of needles similar to full weft yarn 0-0/8-8/0-0/8-8/0-0//.

For the machines with available chain links or electronic drives, any motion is possible, such as placing the yarn around one wale and then connecting to another wale (Figure 2.3) used often in the production of mesh structures, for instance. Another variation is the lapping for covering of larger areas, where the guide moves at each cycle to a large distance in the alternative to the previous direction in order to place a yarn around a larger sector (Figure 2.4)

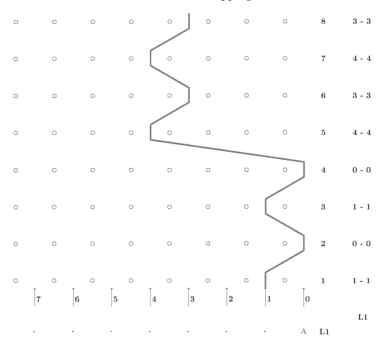

FIGURE 2.3 Lapping for connection of different areas: 1-1/0-0/1-1/0-0/4-4/3-3/4-4/3-3//.

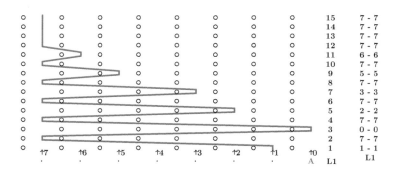

FIGURE 2.4 Lapping for filling some regions lapping movement: 1-1/7-7/0-0/7-7/2-2/7-7/3-3/7-7/5-5/7-7/6-6/7-7/7-7/7-7/7-7//.

2.3   Laps in the same course - pillar stitch

If the guide bar places the yarn over the same needle, and then under the
same needle, it builds a pillar stitch . The pillar stitch can be closed (Figure 2.5 a,b,c,d) if the needle swings-in at the same position (for instance, always at 1) so that closed loops are built(Figure 2.5 e,f,g). In order to produce
a pillar stitch with open loops, the guides have to overlap the needle entirely
from a different side. The pillar stitch can be done as well with alternating
open and closed loops, Figure 2.6

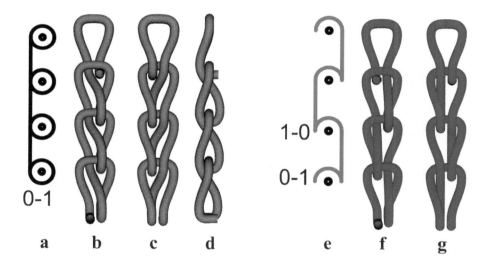

FIGURE 2.5 Pillar stitch with closed and open loops. Closed loops lapping
1-0// ; a) coding and lapping, b) loop back, c) loop face, d) side view; open
loops e) coding and lapping 1-0/0-1// f) loop back g) loop face.

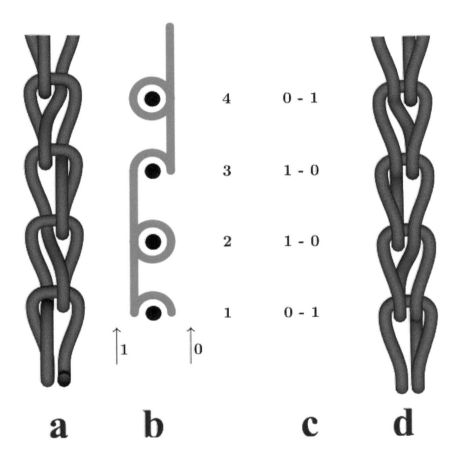

4	0 - 1
3	1 - 0
2	1 - 0
1	0 - 1

a   b   c   d

FIGURE 2.6 Pillar stitch with alternating open and closed loops. L1: 0-1/1-0/1-0/0-1// a) back view, b) lapping movement, c)coding, d) front view.

2.4 Balanced laps in two courses

The most used lapping motions are balanced laps in two courses. For these motions, the guide bar moves under X needles in one direction, overlaps the reached needle, then moves back under X needles and overlaps the needle there. The overlap is usually over one needle, so the laps are coded as X Y lap, where X can be 1,2,3, 4 and so on, and Y is 1, for one needle or 2 for two needles (see Section 2.7). The balanced laps are used normally on two guide bars with opposite motion directions, building asymmetric pattern and for full threaded guide bars then the first number - the underlap X - shows how many yarns of the one guide bar crosses with one yarn from the other guide bar in the underlap (technical face) area. Such fabrics are discussed in Chapter 4.

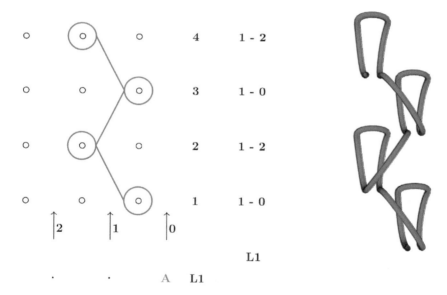

FIGURE 2.7 Closed tricot stitch 1-0/1-2//.

The 1 1 lapping is the simplest of these movements and is named **tricot** lapping, which can be closed (Figure 2.7) and open (Figure 2.8). It produces overlaps in alternate wales at alternate courses. Cord lap is the 2 1 lap, where two yarns will cross (Figure 2.10 for closed and Figure 2.9 for open). Satin lap is a 3 1 lap (Figure 2.11 for closed and Figure 2.12 for open) and velvet is a 4 1 lap. The principal graphical representation of the lapping differs between these only by the numbers of the courses, under which the guide is underlapping and in the real structure or its 3D representation on the length of the underlap.

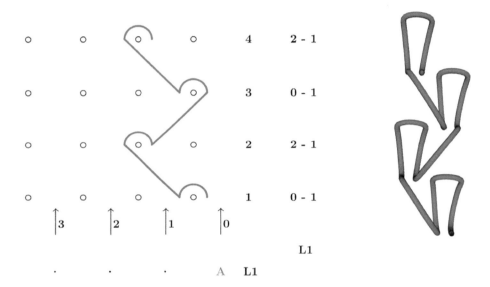

FIGURE 2.8 Open tricot stitch 0-1/2-1//.

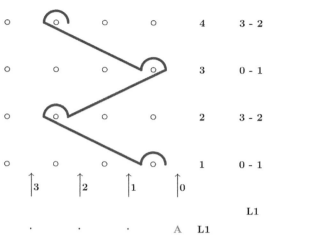

4	3 - 2
3	0 - 1
2	3 - 2
1	0 - 1

↑3 ↑2 ↑1 ↑0

L1

A L1

FIGURE 2.9 Open cord stitch 0-1/3-2//.

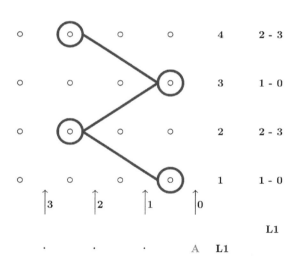

4	2 - 3
3	1 - 0
2	2 - 3
1	1 - 0

↑3 ↑2 ↑1 ↑0

L1

A L1

FIGURE 2.10 Closed cord stitch 1-0/2-3//.

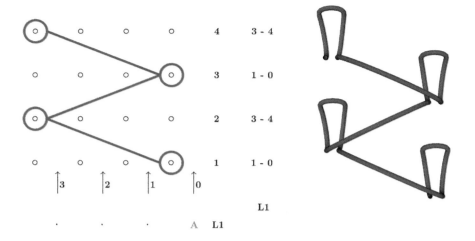

FIGURE 2.11 Closed satin stitch 1-0/3-4//.

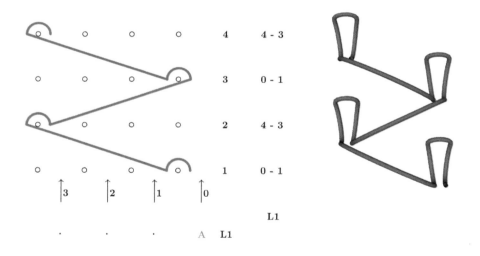

FIGURE 2.12 Open satin stitch 0-1/4-3//.

2.5   Atlas laps

Another way of lapping is the progressive lapping in the same direction in several consecutive courses, normally followed by an identical lapping movement in the opposite direction. Usually, the progressive lapping is in the form of open laps and the change of direction course is in the form of a closed lap (open atlas), but these roles may be reversed [24]. The underlaps on the technical back give the appearance of a sinker loop of a weft knitted structure. The open atlas has practically no underlaps and produced as single guide bar structure is the lightest warp knitted structure. The following figures demonstrate some examples of atlas stitches and the form of yarns. The smallest open atlas with closing turning point is visualised in Figure 2.13 and with the open turning loop in Figure 2.14. The smallest closed atlas with closed turning loop is presented in Figure 2.15 and with open turning loop in Figure 2.16.

One specific parameter of the atlas laps is the number of the courses in which the guide bar laps in one direction. Four courses closed atlas with open turning loop is visualised in Figure 2.17.

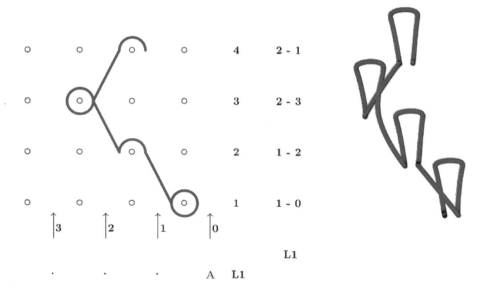

FIGURE 2.13 Open atlas with closed turning points.

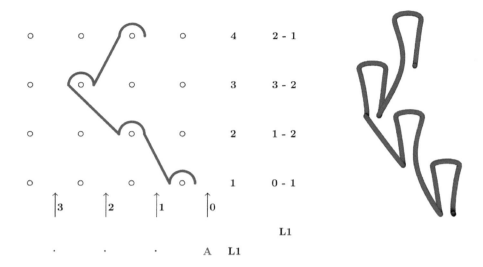

4	2 - 1	
3	3 - 2	
2	1 - 2	
1	0 - 1	

A   L1

FIGURE 2.14 Open atlas with open turning loop.

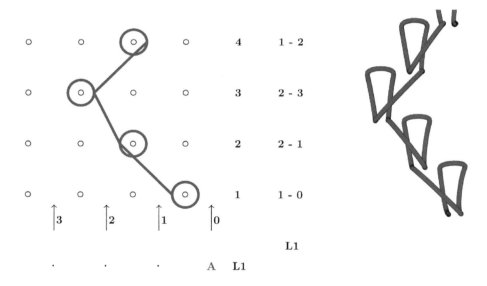

4	1 - 2	
3	2 - 3	
2	2 - 1	
1	1 - 0	

A   L1

FIGURE 2.15 Closed atlas with closed turning points.

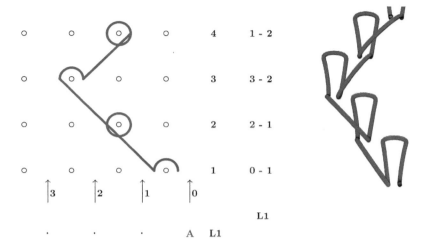

FIGURE 2.16 Closed atlas with open turning loop.

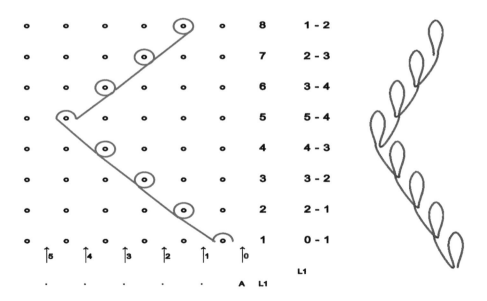

FIGURE 2.17 Four courses closed atlas with open turning loops, lapping: 0-1/2-1/3-2/4-3/5-4/3-4/2-3/1-2//.

2.6 Atlas laps with back laps

The atlas laps can be produced as well with longer underlaps between the loops. In this case these are named back-lapped. Figure 2.18 demonstrates closed backlapped atlas with open turning loop and Figure 2.19 demonstrates the different possible combinations. During the back lapping longer underlaps are placed, which limits the elasticity of the fabrics and gives a smooth surface on the technical face.

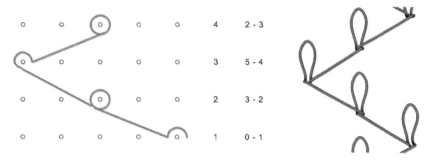

FIGURE 2.18 Closed backlapped atlas : 0-1/3-2/5-4/2-3//.

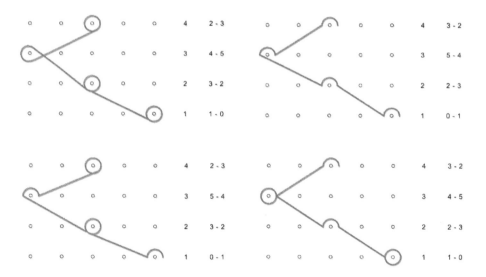

FIGURE 2.19 Different backlapped atlas lappings.

2.7 Laps over two needles

In most warp knitted structures, the guide places yarn over one needle and only one loop per knitting cycle is built from one yarn. The yarn length supplied and placed over this one needle and has to be enough for the loop formation process. Overlapping over two needles is as well possible - with proper run-in settings and good elasticity of the yarn it can be placed over two needles and in this way two adjacent loops in the same course can be done at the same cycle. This type of pattern is named "Köpper" in the German language, which means "twill", in analogy to the twill structure in weaving, where the warp and weft yarns float over more yarns. Theoretically, all previously described patterns could be produced with proper material and proper run-in settings with lapping over two needles, some with more difficulties than others, practically only few of these are used. Figure 2.20 a demonstrates the lapping movement for open lock stitch over two needles, closed lock stitch over two in Figure 2.20 b, open tricot in Figure 2.20 c and closed tricot in Figure 2.20 d. The corresponding idealized 3D images are visualized in Figure 2.21. Closed and open laps and atlas laps over two needles are demonstrated in Figure 2.22

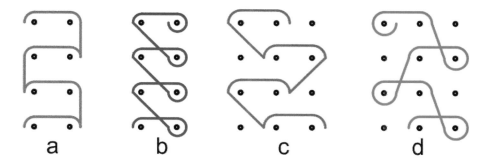

FIGURE 2.20 Main laps over two needles a)open lock stitch 2-0/0-2// b) closed lock stitch 2-0// c) open tricot 0-2/3-1// d) closed tricot 2-0//1-3//.

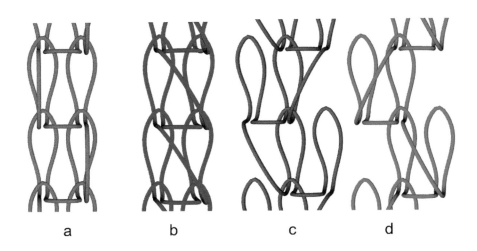

FIGURE 2.21 3D Visualisation of the main laps over two needles a)open lock stitch 2-0/0-2// b) closed lock stitch 2-0// c) open tricot 0-2/3-1// d) closed tricot 2-0//1-3//.

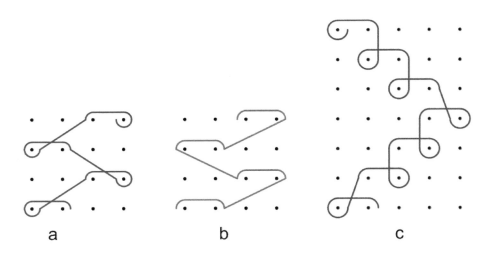

FIGURE 2.22 Laps over two needles with larger movements a) closed cord stitch 2-4/2-0// b) open cord stitch 4-2/0-2// c) atlas stitch.

2.8 Special lapping movements

Mainly for some special effects in crochet knitting are used single bars, which are working at five or more tempi and provide five or more motion steps in the time of knitting one loop with the other guide bars. Such effects are used to create multiple circles around a needle for decoration purposes. Figure 2.23 demonstrates two such motions as extension to the closed loop (2.23 a), such one based on four or five tempi (2.23 b) and six or seven tempi (2.23 c). Normally such laps can be produced with special devices on the guides and can be found on machines of the companies CO.ME.TA. S.r.l. `http://www.cometaitaly.it` and Comez Knitting Systems `www.comez.com`.

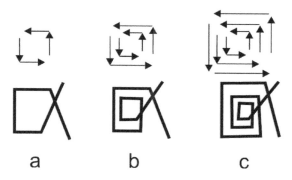

a   b   c

FIGURE 2.23 Special laps for building decorative elements.

2.9 Conclusions

This chapter presented an overview of the main lapping motions for guide bars of warp knitting machines. The lapping motions give the foundation of warp knitting patterning as the pattern consists of several laps in parallel. Several knitting people have drawn the pattern before professional warp knitting CAD systems were invented using vector based graphic software as AutoCAD, CorelDraw, Drawing tools in Microsoft Office, etc., creating the graphical images of single lapping and then copying and combining these in parallel. Today modern warp knitting software normally has a library with the main lapping motions and these can be directly selected there (see Chapter 13).

Part II

Loop based single face structures

3

Single guide bar fabrics

3.1 Introduction

Single guide bar fabrics are the simplest warp knitted fabrics, produced from one guide bar, usually by full threading.

All yarns in such fabrics have the same orientation, and the forces in the loop heads are often come in equilibrium after deformation of the structure, so the loops can have different angles. Single guide bar fabrics can be interpreted as the basic element, which is used for the design of more complex structures and multiple bar structures. The multi bar structures are a combination of several single bar structures, but not all pattern and threading of single guide bars produce stable structures. This chapter covers only those single bar structures, which are topologically stable - it means they do not destroy their intersections, but they can change their geometrical form - and exist as a separated fabrics. These are used in the real world in specific applications, where the form stability is not important (for instance meshes for vegetables or other products) or where even the flexibility and the porosity are very important - as for instance for medical applications.

3.2 Pillar stitch chains

Pillar stitches are built when the yarn guide produces lapping around one and the same needle. Such motion leads to production of vertical courses, which **are not connected into fabrics**. These are used mainly often as a ropes, or as a basic element in meshes and as a ground for structures, connected by weft yarns or other loops. All three types of the pillar stitches can be used - the open, closed and open-closed (Figure 3.1).

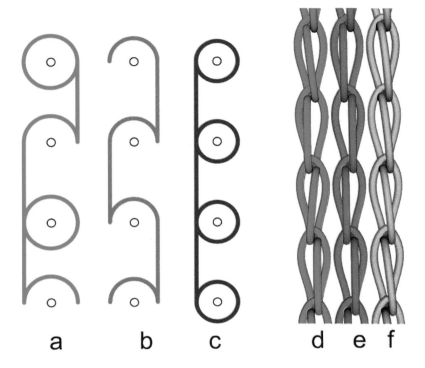

a   b   c   d e f

FIGURE 3.1 Pillar stitches as loop chains a) and d) open - closed; b) and e) open, c) and f) closed.

3.3   Tricot stitch fabrics

Tricot fabrics are the simplest plain fabrics, where the individual chains are connected at every course. The example structure with closed lap is demonstrated in Figure 3.2 as idealized simulation Figure 3.2 b), with adjusted vertical orientation of the loops Figure 3.2 c) and as a microscopic image of real structure (Figure 3.3). The open tricot is visualized in Figure 3.4.

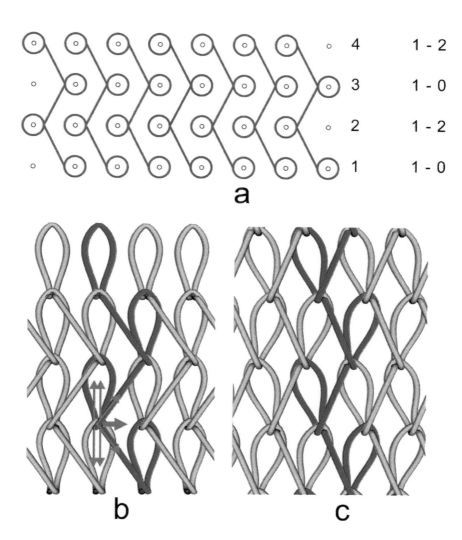

FIGURE 3.2 Closed tricot fabrics a) lapping movement, b) idealized 3D simulation with vertical loops c) 3D simulation with adjusted loop orientation. Lapping movement GB 1: 1-0/1-2// ; Threading: BBBABBB.

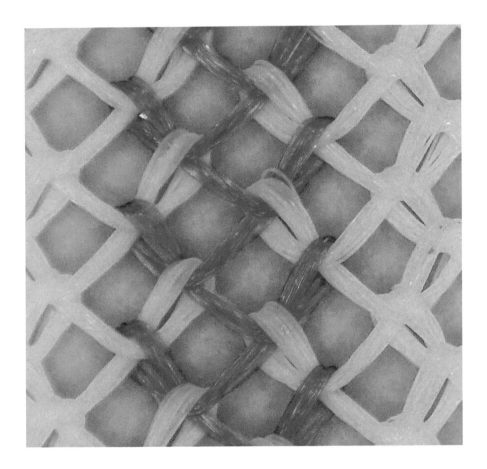

FIGURE 3.3 Closed microscopic image of fabric from [23], p.21.

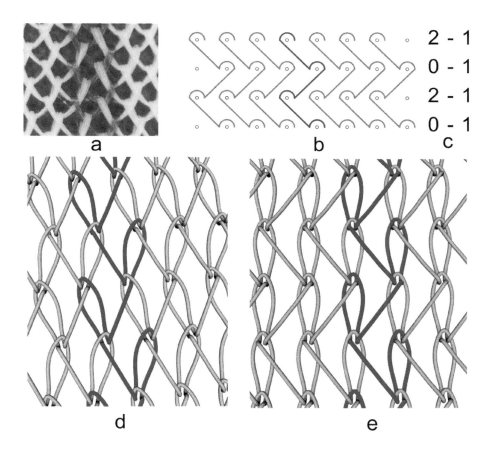

FIGURE 3.4 Closed tricot. a) microscopic image of fabric from [25], b) lapping diagram, c) numerical coding, the repeat is 0-1/2-1// ,d) simulated 3D image with loop orientation correction e) simulated 3D image with vertical loops.

3.4   Cord stitch fabrics

Cord stitch fabrics at full threading can be recognized through the underlap behind one loop. The configuration with closed loops in simulated and photographic images is represented in Figure 3.5. The underlaps determine the optics of the technical face side Figure 3.5 f), while on the technical back the loop faces are visible Figure 3.5 g). The loops in the relaxed stay are more declined to the horizontal side in comparison to the loops of the tricot stitch, because of the larger paths to left and right. Open cord stitch fabrics can be recognized under the microscope by looking on the loop legs (Figure 3.6). The loop legs do not cross in the open loops (Figure 3.6 a) and are oriented one over the other in a vertical direction, which makes their loops a little bit narrower than the loop of the closed cord stitch fabrics (Figure 3.6 b), where the crossed legs are placed at one vertical position in the loop head. The cord stitch can be used to create fabrics with partial threading, where every other guide receives yarns. Figure 3.7 demonstrates such a sample with threading A.B.A.A., where the yarn type B is used only for better visualization. The distance between two courses in this structure is twice as large as the needle spacing t and after the sample is taken out of the machine it is classified as a "tricot" fabrics structure, because there is no topological difference in the yarn interlaces between the tricot and the half threaded cord stitch. This technique is used if a machine with another fineness is not available, and the yarns are still not so thick, so that they can be processed with the finer needles of the finer machine.

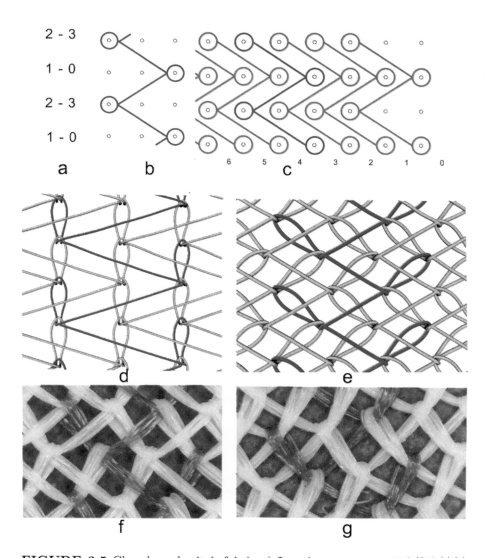

FIGURE 3.5 Closed cord stitch fabric a) Lapping movement 1-0/2-3// b) graphical representation, c) technical drawing of the full threaded fabrics with one different yarn in the middle; d) simulated fabrics with corrected loop orientation - technical face [13]; f) microscopic image of fabrics from [25] technical face and g) technical back.

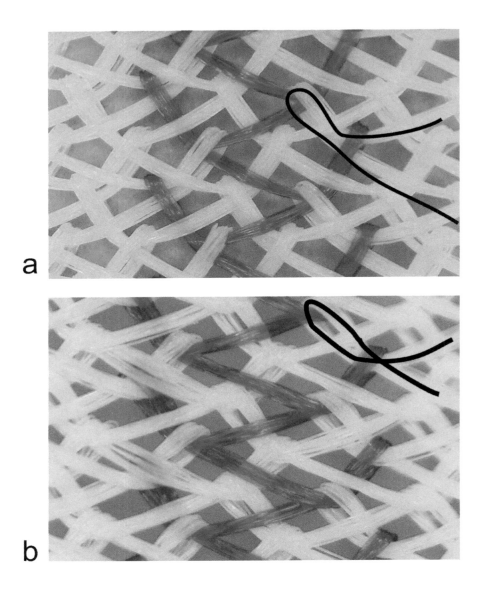

FIGURE 3.6 Open and closed cord stitch, from [23], p. 27 and p. 28.

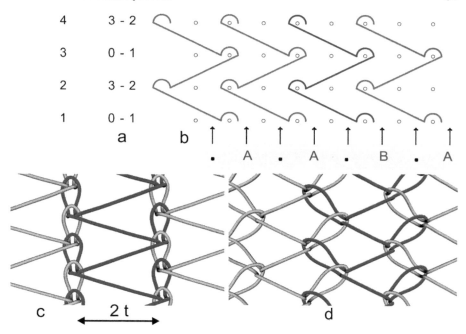

FIGURE 3.7 Open cord stitch with partial threading A.B. (one in - one out), which produces fabrics with double larger distance between the courses and appearance as full threaded tricot stitch. a) coding, b) technical drawing with the threading, c) simulated structure with vertical loops d) simulated structure with deformed loops, closer to reality[13].

3.5   Satin and velvet fabrics

For the structures, where longer floatings of the yarns are needed, the satin (under 3 over 1) and velvet (4 - 1 or under 4 over 1) lapping in full threading are used. These structure have more shine and a smooth technical face because of the longer underlaps. Figure 3.8 demonstrates full threaded open satin fabrics and Figure 3.9 - closed velvet fabrics. Using partial threading - 1 in 2 out or (A..) for satin or 1 in 3 out (A...) for velvet can be used as well for production of tricot fabrics with lower density. Using velvet with 1 in 1 out (A.) leads to a cord fabrics with lower density. Figure 3.10 demonstrates this "transformation" of the velvet to cord fabrics, as the upper figure has full threading, but with very thin yarns in the guides, which becomes empty for the partial threading, represented down.

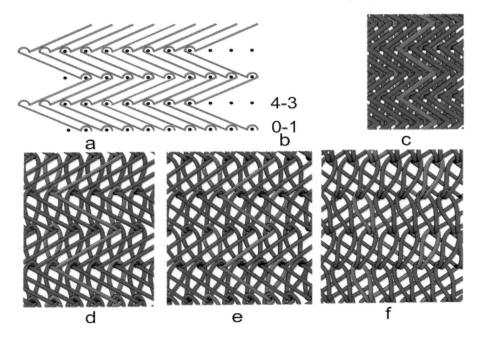

4-3

0-1

FIGURE 3.8 Open satin fabrics with full threading a) graphical representation, b) numerical coding of the lapping, c) technical front simulated with
thick yarn closer to the reality visualisation d) and e) technical back without
and with loop orientation correction, f) technical front.

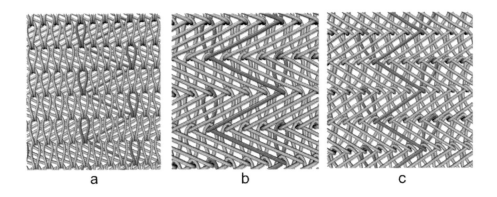

FIGURE 3.9 Velvet fabrics with full threading a) technical front - loop
side; b) technical back with visible underlaps under three courses, idealized
simulation; c) with loop orientation correction.

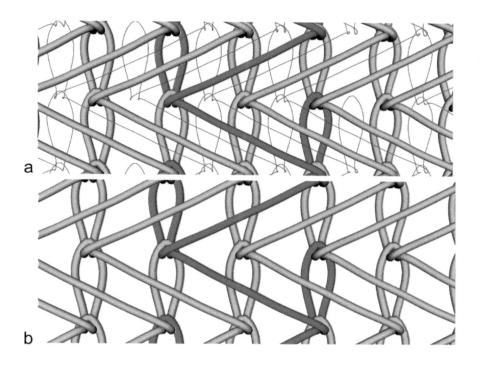

FIGURE 3.10 Partial threading (A.) and velvet lap leads to cord lap fabrics with lower density. a) velvet with full threading, but yarns, which will be removed, are thin, b) resulting cord stitch fabrics produced with velvet lapping and partial (A.) threading.

3.6   Atlas fabrics

The atlas lapping can be used for the production of the lightest fabrics. The technical back of the open atlas (Figure 3.11) looks very similar to horizontal stripes of weft knitted fabrics, where the stripe width depends on the number of the courses of the atlas (Figure 3.12). These structures have the lowest weight per unit surface as they are built with the shortest yarn length per course, because there are practically no underloops there. Closed atlas fabrics are similar, but built of closed loops (Figure 3.13 and 3.14). The loops of single guide bar atlas fabrics do not remain vertical and are tilted to one or other side because of the assymetric loading from the yarns of their previous and following loops.

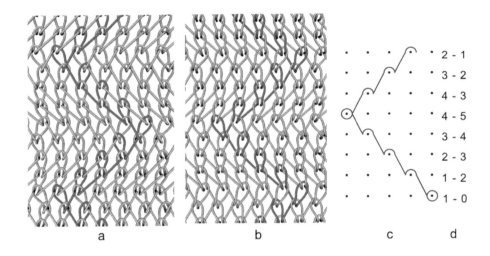

a   b   c   d

FIGURE 3.11 Open four rows atlas fabrics with full threading a) technical front, b) technical back, c) lapping movement, d) coding.

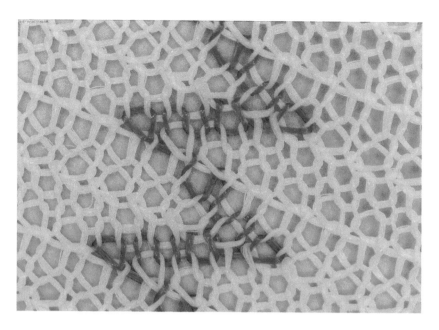

FIGURE 3.12 Open four rows atlas fabrics, microscopic image of sample from [23], p. 33.

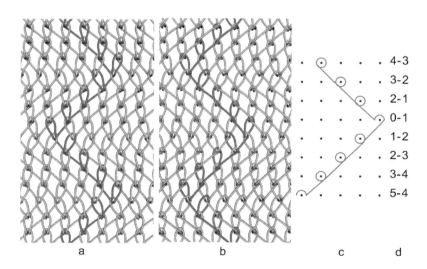

a   b   c   d

FIGURE 3.13 Closed four rows atlas fabrics with open turning point, a) front side, b) technical back, c) lapping diagram, d) coding.

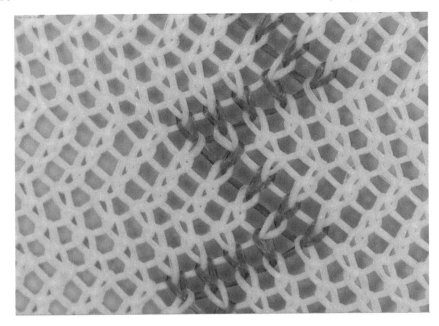

FIGURE 3.14 Closed four rows atlas fabrics, with closed turning loops, microscopic image of sample from [23], p. 34.

3.7 Back lapped atlas with full and partial threading

The lapping diagram of back lapped atlas fabrics with full threading is presented on Figure 3.15. The longer underlapp under two needles reduce the elasticity of the fabrics. Simulated fabrics with full trheading is shown on Figure 3.16a). Figure 3.16b) demonstrate the transition from back lapped atlas to normal atlas, if each another yarn becomes completely removed (Figure 3.16c). This structure may be recognized as an normal atlas from machine with double coarser gauge, but the experienced knitters would normally recognize the half threading based on the loop size - coarser machine will cause larger loops. The correct documentation of the sample 3.16c) require the partial threading and back lapped motion.

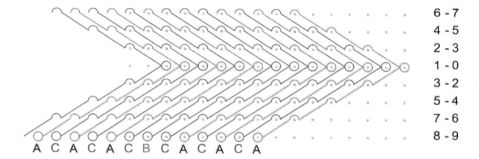

FIGURE 3.15 Backlapped open atlas full threading. If the C-type yarns are removed, the partial threading leads to normal atlas with double lower density.

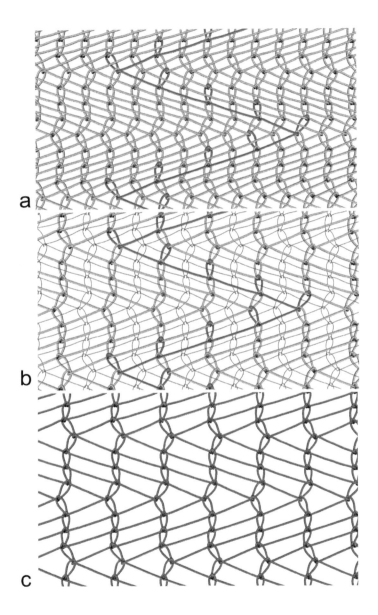

FIGURE 3.16 a) Open backlapped atlas in full threading, technical back
b) each other yarn is marked thinner and demonstrate the effect of partial
threading c) same pattern with each other yarn removed, and - is normal
atlas with half the number of courses per unit width.

3.8 Fabrics with lapping over two needles

When one needle places yarn over two needles, two loops are built in the same course (row) (Figure 3.17 a.), and if full threading of the guide bar is used, plated loops (with two yarns each) are built (Figure 3.17 b.).

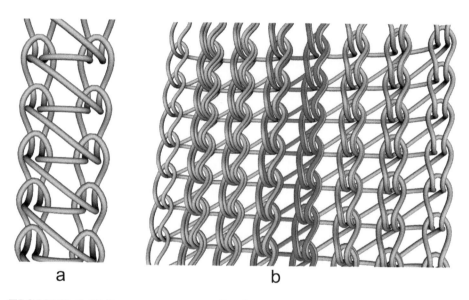

a   b

FIGURE 3.17 Laps over two needles build two loops in a course, here an example of lock stitch 2-0//, full threading.

The two loops per course allow creation of fabrics with partial threading, as visualised in Figure 3.18 for tricot structure over two and Figure 3.19 for atlas based structure over two needles. Due to the higher technological difficulties and sensible knitting process, such structures are seldom used.

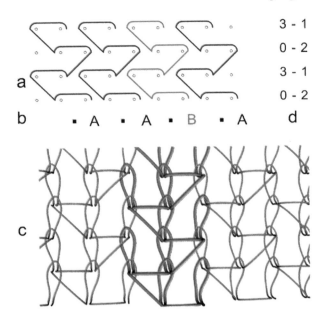

FIGURE 3.18 Laps over two needles build two loops in a course, here an example of lock stitch 2-0//, full threading.

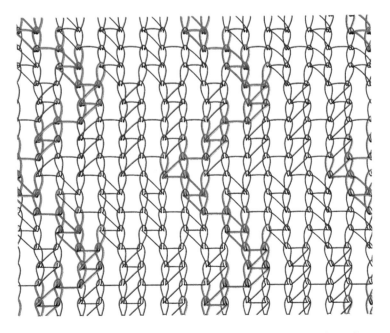

FIGURE 3.19 Alas lapping over two needles with partial threading.

3.9 Combined stitches

All basic stitch types can be combined in a sequence for the building of more complex lapping. Such combinations are used normally for fabrics of two or more bars (multibar technique), but in some cases they can be applied as well on single guide bar structures. One such example is the combination of lock stitch and tricot for the building of a simple mesh structure (Figure 3.20). The length of the part with lock-stitch determines the length of the mesh elements (Figure 3.21).

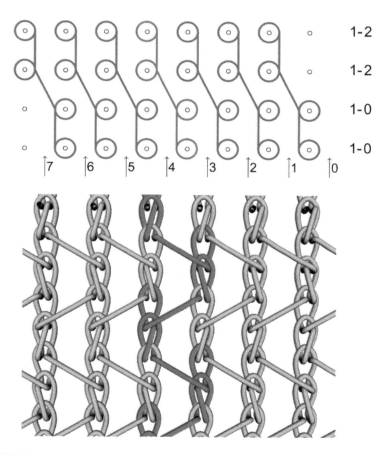

FIGURE 3.20 Mesh structure, created by combination of tricot lapping for connecting the mesh elements and lock stitch for the vertical elements lapping: 1-0/1-0/1-2/1-2//.

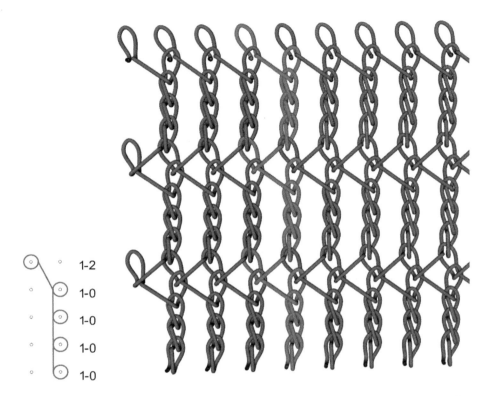

FIGURE 3.21 Mesh structure, created by combination of tricot lapping for connecting the mesh elements and lock stitch for the vertical elements lapping: 1-0/1-0/1-0/1-0/1-2//.

3.10 Conclusions

Tricot, cord, velvet, satin and atlas laps applied on single guide bar with full threading leads to production of warp knitted fabrics. The lock stitch produces only vertical chains, which are not connected together into wider fabrics. All laps over two needles (including lock stitch 0-2/0-2//) produce fabrics; they can be used as well with full and half threading. The laps with longer underlaps (cord, velvet, satin) and the back lapped atlas can be applied with partial threading of the guide bar for production of fabrics with lower density on the same machine. Single guide bar fabrics are very light structures, whose loops change their vertical orientation during the relaxation process.

4

Two full threaded guide bars with tricot lapping

4.1 Introduction

When using two (or more) guide bars, plated loops of two (or more) yarns are built. This chapter presents rules for orientation of the loops and discusses the possible combinations of laps and colour patterning on the basis of tricot lapping. Combinations of other lappings are given in Chapter 5.

4.2 Plated loops

The structures with two fully threaded guide bars consist of **plated** loops. Plated loops are built of two yarns (Figure 4.1), where one is in most cases placed over the next one. The exact configuration of the two loops depends on the yarn properties and the pattern. If the two yarns have significantly different lengths or bending stiffness, which can happens for elastic, smart or technical textiles, (hyper-elastic material, conductive yarns, stiff carbon or glass yarns, wires) - the loop form of each of these materials will be different. The pattern in this chapter covers the main rules for normal, standard fabrics, where both of the guide bars have the same or similar material.

FIGURE 4.1 Plated loop consists of two yarns.

4.3   Positions of the loops and underlaps

The position of the **underlaps** in warp knitted structures of two (or more) guide bars is always exactly determined and depends on the **relative position of the guide bars** only. The guide bar, which is far away from the needles, places its underlaps in the most outer side of the fabrics and these remain visible. If the underlaps are in the same direction (equal lapping) and there is a difference in the run in, it can happen that the yarns of second (or other) guide bars become visible, but for the case of counter-lapping, where the underlaps are crossign together, always the first guide bar builds the visible yarn pieces. This is demonstrated in (Figure 4.2), where both samples (a-b-c) and (d-e-f) represent the same double tricot structure where GB1: 1-0/1-2//; GB2: 1-2/1-0// with full threading. At the first sample one yarn with a different colour is placed on the GB2. Neither loops nor the underlaps of this yarn (and all other yarns on the GB2) become fully visible, only small parts of these yarns can be seen between the underlaps and loops of the first guide bar. On the contrary, one colour yarn of guide bar GB1 (and all other yarns of this guide bar) are visible on both sides and determine the colour appearance of the fabrics.

While the underlap position of each yarn depends always from the (number and respectively the) position of the guide bar, the position of the yarns in the plated loops depends on more factors and cannot always be definitely determined. The orientation of the yarns over the loops depends on the placement of the yarns over the needle during the overlapping process and can change if the yarns can slide one over another during the subsequent knitting process. The yarn, which is in the most bottom position over the needle will determine the technical face of the fabrics with the loops. In the case of Figure 4.3 a this is the blue yarn and in the case of Figure 4.3 b this is the red yarn. Which of these situation will happen for each loop depends on the three dimensional

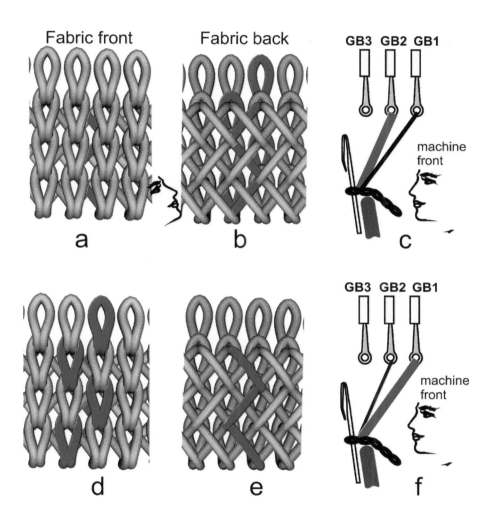

FIGURE 4.2 Closed tricot-tricot pattern with counter-lapping. One colour yarn in GB2 (c) is less visible on both sides, where the colour yarn from the first guide bar - GB1 (f) makes visible loops (d) and visible underloops (e).

geometry of the yarns and the needle, which is determined by the lapping motion and the guide bar positions and can be influenced from the yarn tension and the run-in, respectively.

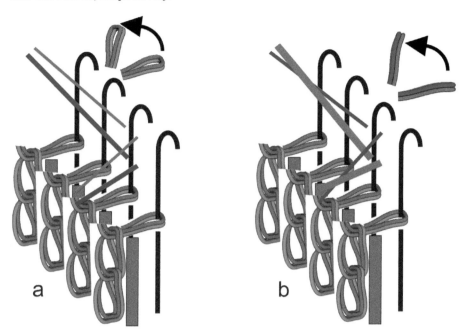

FIGURE 4.3 Plating loops - the bottom yarn on the needle after the overlapping and swing out motion will remain visible. Depending on the yarn tension, lapping motion and the guide bar position this can be a) the blue yarn and b) the red yarn.

Figure 4.4 demonstrates the loop building process of two plated loops [18], p. 81. During the knitting process both guide bars swing in and overlap their yarns over the single needles (Figure 4.4,a and b). During the swing-out motion (Figure 4.4,c) the yarns of the front bar GB1 are the first to strike the needle beards or shaft. The yarns of the second bar do not contact the needle until later (Figure 4.4,d). If both overlaps are made in the **same direction**, the yarn of the front bar (GB1)prevents the yarn of the second bar from the sliding down along the needle. "Thus, providing that these positions are retained through subsequent knitting of the course, the front bar threads, because they are lower on the needle stems, will be plated on the face of the fabric"[18], (Figure 4.4,e, f, g). If the **overlaps** are made in **opposite direction**, the yarns will be crossed on the needle stems. The tendency is then that one half of each loop shows the yarn supplied by the front guide bar and to the other half of the loop to show the yarn, supplied by the second

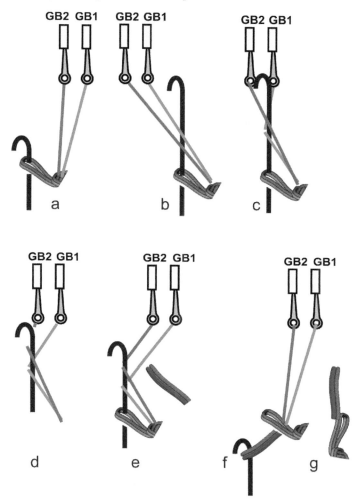

FIGURE 4.4 Plating loops in the most common case.

guide bar. The yarn from the front guide bar (GB1) will be more prominent on the face side of the fabrics, because it is closely impinged upon the needle stem during the lapping action. If the warp tensions and the heights of the guide bars are correctly adjusted, the front bar threads can be caused to plate entirely on the face [18].

Larger differences in the previous positions of the guide bars (underlap) influence additionally the geometry in the knitting area. As demonstrated in Figure 4.5 this can cause the blue yarn, which made shorter motion, to remain outside.

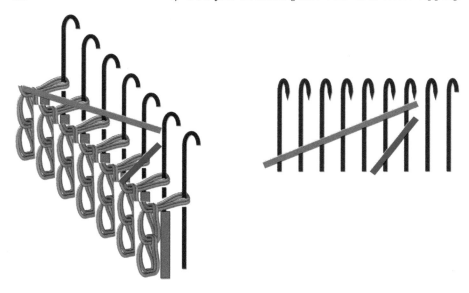

FIGURE 4.5 Large difference in the underlap changes the position of the yarn.

4.4   Vertical orientation of plated loops

The warp knitting machines work normally with multiple guide bars. If the two bars are moving in the same direction during the **under**lapping motion, the lapping is *equal*. The loops normally are placed in this case sloped under some angle, coming from the direction of the yarns that are underlapped.

If the two bars have **under**lapping movements in opposite directions - the lapping is counter-lapping (in German: gegenlegig). In this case there is no tension on one side of the loop and the loops remain straight vertically(Figure 4.6a). The lapping over the needles (overlap) can be in simultaneously or opposite direction, but this does not influence the vertical position of the loops. If the lapping in both steps - overlap and underlap are performed with motions in opposite direction - then the patterns are named "doubled"- as "double tricot, "Double - cord" etc. [18], [23]. p. 15.

In the case of equal lapping the underlaps will have the same orientation; the sum of the tensile forces in all underlaps is a resulting force, which moves the loop head to one direction (Figure 4.7c and c) and causes side inclination of the loops (Figure 4.6c). The underlaps in opposite directions, placed by counter-lapping, compensate the force mutually and result in a zero horizontal component of the resulting (Figure 4.7a and b) force, and the loops remain vertical without side inclination (Figure 4.6a).

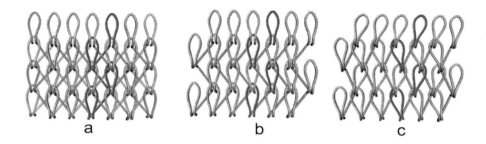

FIGURE 4.6 Vertical orientation of the loops. a) loops of two guide bars with counter-lapping remain vertical; b) idealized geometry of two identical (or with equal lapping) produced guide bars, c) side inclination of the loops.

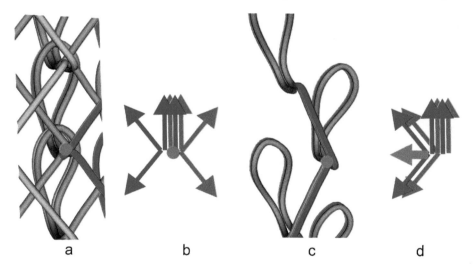

FIGURE 4.7 Forces in plated loops a) underlaps orientation at counter-lapping b) forces from the current yarns c) loops for equal lapping, d) forces in the loop head at equal lapping. The resulting force has a significant horizontal component, which moves the loop centre and leads to loop declination.

4.5   Tricot-tricot combinations

Tricot lapping consists of two cycles, each of which can have closed or open loops, which means that four tricot lappings are possible. Two guide bars working with tricot-tricot arrangement can produce mathematically sixteen different combinations of pattern (Figure 4.8) in case of equal lapping and additionally sixteen configurations of pattern in the case of counter-lapping (Figure 4.9). Although only few of these configurations are used in reality, a precise notation of the lapping is required, if such samples have to be reproduced exactly. The naming of "tricot-tricot" is fairly not enough, as some people at the beginning of their contact with warp knitting are thinking. The differences between these structures can be recognized only after careful analysis of the loop legs under a microscope, as Figure 4.10 demonstrates, from the initial view of the geometrical representations they all look very similar. Figure 4.11 gives the impression of the real structure of closed tricot-tricot structure in counter-lapping in both face and back side. The loop heads in the photo demonstrate that the plated loops for multifilament yarns with minimal twist are a complex union of filaments, and their position is not always exactly the same. For this reason, it is often said that the the yarn of the one guide bar *"dominates"* the other, which means that this does not have to be true for all loops. The arrangement of the underlaps is always identical and depends only on the guide bar position.

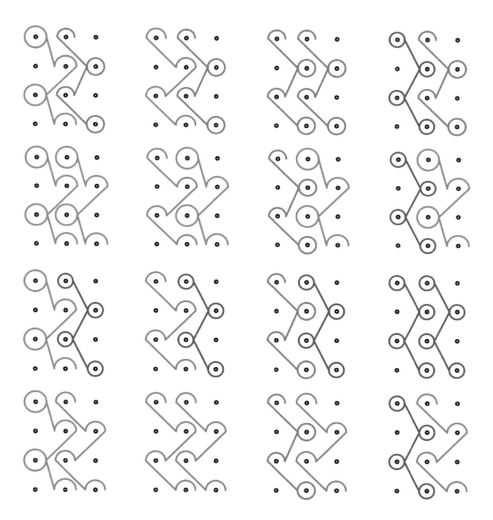

FIGURE 4.8 All sixteen possible tricot-tricot combinations with equal lapping.

FIGURE 4.9 All sixteen possible tricot-tricot combinations with counter-lapping.

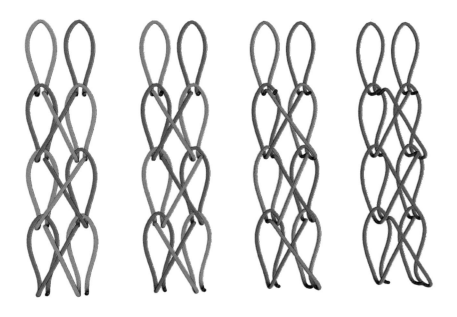

FIGURE 4.10 3D view of four examples of the combinations of tricot-tricot with counter-lapping drawn from Figure 4.9.

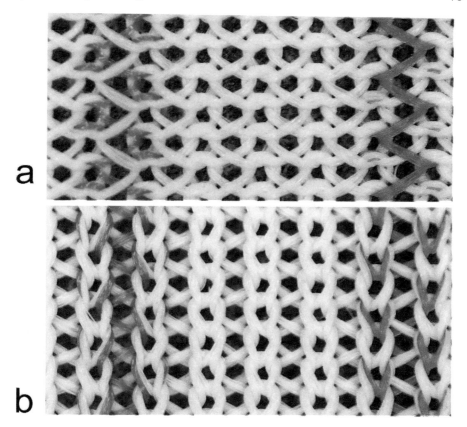

FIGURE 4.11 Closed tricot-tricot lapping in counter-lapping, photographed from [23], p. 26.

4.6 Colour patterning

The dominance of the *first* guide bar over others in the plated loops is used for selecting and patterning with colour yarns. If the first guide bar (GB1) is fully threaded with blue yarns (Figure 4.12) and the second (GB2) with red yarns, then for counter-lapped, closed double tricot mainly the blue yarns will be visible on the face side (Figure 4.13b) and again the blue yarns will dominate with its underlaps over all other overlaps yarns on the back side (Figure 4.13a).

Using some colour sequence, for instance four red - four blue yarns in the first guide bar (Figure 4.14a) allows the creation of vertical stripes on both face (Figure 4.14b) and back side (Figure 4.14c). The threading of the second guide bar in this case is kept identical in order to have a clean optical appearance. The yarns of the second guide bar become more or less visible between the loops and behind the underlaps on both sides, depending on the yarn thickness, machine fineness and the loop height. Shifting the second guide bar (GB2) on two positions (Figure 4.15a) does not change the positions of the plated loops and the visible underlaps of the first guide, but creates less clear, unpurified stripes between the clean red and clean blue stripes. Complete opposite threading of the second guide bar, achieved by the shift of four needles (Figure 4.15b) makes complete stripes with mixed optics - one colour of the

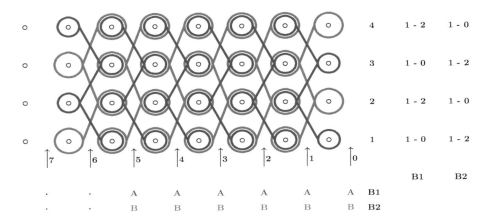

FIGURE 4.12 Lapping movement GB 1: 1-0/1-2/1-0/1-2// ; GB 2: 1-0/1-2/1-0/1-2// ; Threading: GB 1: A A A A A A ; GB 2: B B B B B B.

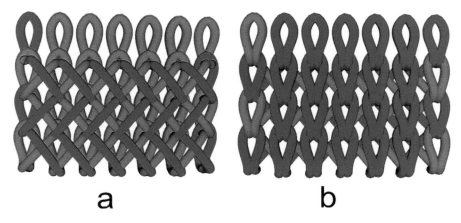

FIGURE 4.13 Double tricot simulation of the pattern of Figure 4.12 a) back side, b) loop side.

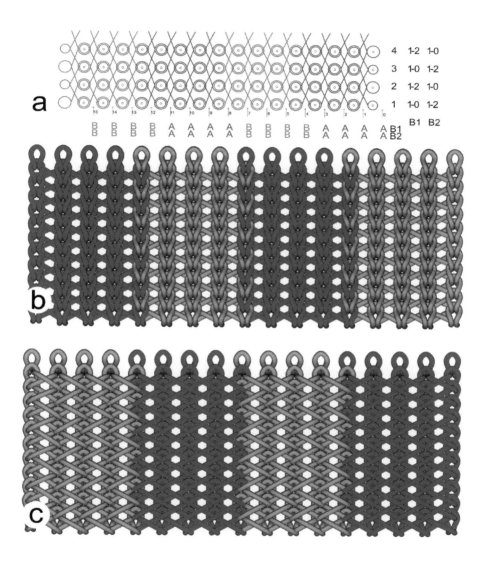

FIGURE 4.14 Pattern with double tricot for vertical stripes GB 1: 1-0/1-2//; GB 2: 1-0/1-2//; Threading: GB 1: A A A A B B B B A A A A B B B B ; GB 2: A A A A B B B B A A A A B B B B.

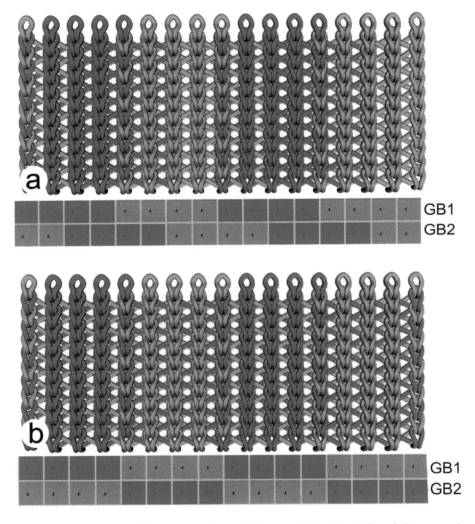

FIGURE 4.15 Double tricot stripes of Figure 4.14 with shifts of the second guide bar a) two needles to the left b) four needles to the left.

loops and the opposite colour of the eventually visible underlaps. These effects can be used for creating areas of clean and mixing colour between the clean colours of the yarns.

4.7 Conclusions

This chapter demonstrated the large number of variations possible for two full threaded guide bars with tricot lapping. Knowledge about the visibility of the yarns in the plated loops and the arrangement of the underlaps is important for the design of warp knitted samples, as it determine the optical appearance and complete surface properties such as softness, abrasion resistance, pilling, etc.

5

Fabrics with full threading

5.1 Introduction

The main principles about the visibility of the underlaps and vertical positions of the loops, described in the previous chapter for tricot-tricot structures, remain valid for all possible combinations of patterns with full threading with two or more guide bars. This chapter demonstrates fabrics, created as combinations of different laps.

5.2 Terminological issues

The fabrics, produced with two guides bars with different laps are named for their laps, for instance, tricot-cord structure. The first lap is that of the first guide bar (GB1) and the second - that of the second guide bar (GB2). This may sound trivial, but in the past the guide bars on some warp knitting machines were numbered in the opposite way. For instance, in the book of Rogler and Humboldt [23], if guide bar 1 produced tricot and guide bar 2 - cord, then the structure is named cord-tricot, because the cord stitch is the visible one. Unified numbering of the guide bars is regulated by DIN ISO 10223:2005, so the reader should be careful about the guide bar positions when reading books and instructions from past years.

5.3 Cord stitch

The cord stitches can be, in the same way as the tricot stitches, open, closed and half open and half closed. Combining these variations with counter- and equal lapping, there are 32 combinations of pattern, but in practice only the open 5.1 a and closed 5.1 b double cord structures are used.

Closed double cord stitch fabrics with full threading and counter-lapping is visualised in Figure 5.2 and the microscopic image of the real structure is

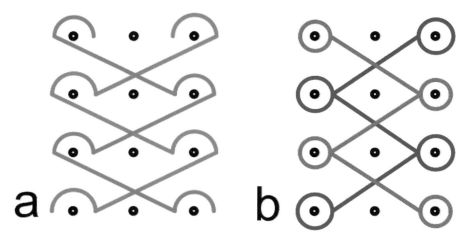

FIGURE 5.1 Double cord stitch, open and closed, with counter-lapping a) open GB 1: 0-1/3-2// ; GB 2: 3-2/0-1// b) closed GB 1: 1-0/2-3// ; GB 2: 2-3/1-0//.

given in Figure 5.3. The first guide bar (GB1) dominates the technical back side and appears as a loop on the front side, while the yarns of the second guide bar become visible mainly between the loops of the front side.

The fabrics with the double open cord stitch with counter-lapping have a similar appearance in the simulation (Figure 5.4) and photos (Figure 5.5). More careful analysis of the photo leads to the observation that the blue yarns, which are working on the second guide bar and are not visible on the technical back, plates over the loops of the first guide bar and become visible on the loop (front) side.

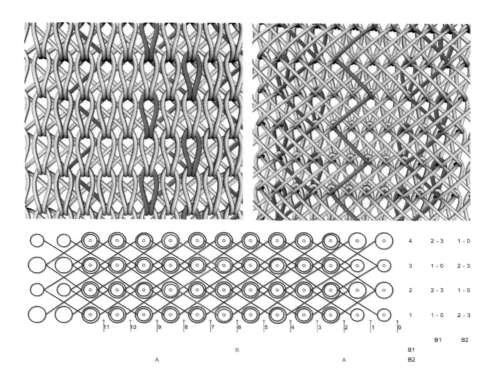

FIGURE 5.2 Double cord stitch, closed, with counter lapping GB 1: 1-0/2-3/1-0/2-3// ; GB 2: 1-0/2-3/1-0/2-3//.

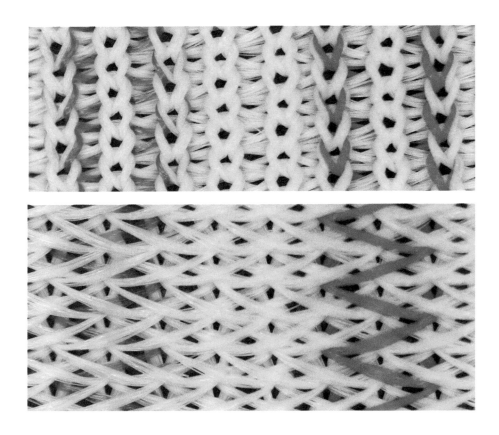

FIGURE 5.3 Photos of samples with the cording from 5.2, microscopic images of samples from [23], p. 29.

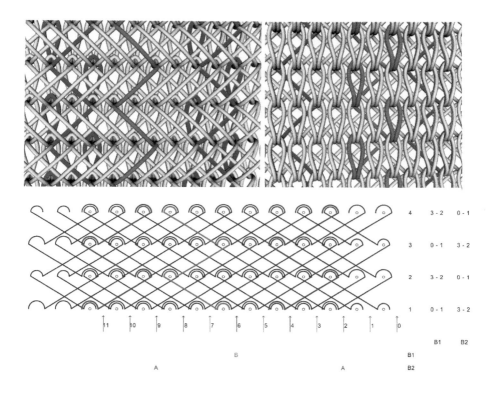

FIGURE 5.4 Double cord stitch, open, with counter-lapping GB 1: 0-1/3-2/0-1/3-2// ; GB 2: 0-1/3-2/0-1/3-2//.

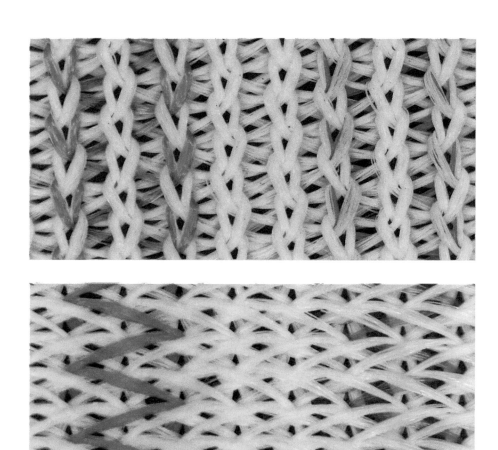

FIGURE 5.5 Photos of the double open cord with counter-lapping sample
with the coding from 5.4, microscopic images of sample of [23], p. 29.

5.4 Cord-Tricot Combinations

The cord and tricot laps as a combination create fabrics with low elongation, because of the longer underlaps of the cord stitch. If the first guide bar places a cord stitch, the longer floats will be visible on the back side (Figure 5.7a). The pattern with equal lapping (Figure 5.6a) leads to significant side declination of the loops and rotation of the loop heads (Figure 5.8). The cord-tricot structure with counter lapping (Figure 5.6b) with polyamide or viscose yarns is often named Charmeuse, and the structure, the Locknit structure. The longer underlaps of first guide bar with the cord stitch determines the back side (Figure 5.10a and Figure 5.9a) and the same material dominates as well in the plated loops on the front side (Figure 5.10b and Figure 5.9b).

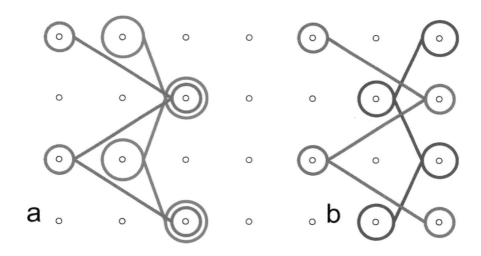

FIGURE 5.6 Cord-tricot combination based pattern. a)with equal lapping, b) with counter-lapping, known as locknit fabric construction.

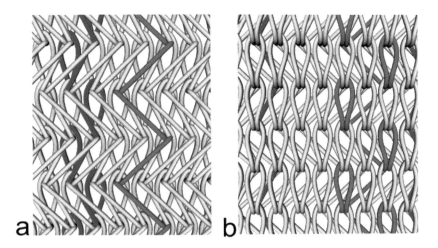

FIGURE 5.7 Cord tricot with equal lapping simulation a)back with visible longer underlaps of the cord stitch b) front - the loops of the cord stitch dominate.

FIGURE 5.8 Cord tricot with equal lapping photo a) loop side b) back side.

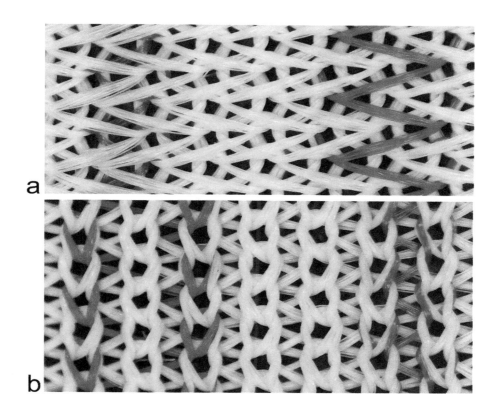

FIGURE 5.9 Cord-tricot with counter lapping a) back side, b) front side.

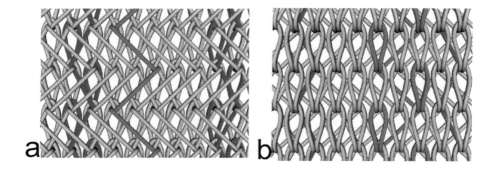

FIGURE 5.10 Cord-tricot with counter lapping simulation a) back side b) front side.

5.5   Tricot-Cord Combinations

The change of the pattern of the two guide bars changes the appearance of the fabrics. If the first guide bar (GB1) applies tricot stitch, its yarns with the shorter underlaps will be the most visible on the both sides of the fabrics. The longer underlaps of the cord stitch will be integrated inside of the structure as visible on the simulation (Figure 5.11) and the photo of the sample (Figure 5.12). Analogously to the cord-tricot, the pattern in the fabrics with equal lapping, the loops change their vertical orientation significantly (Figure 5.13b). In this case the shorter tricot underlaps are not really able to fix and keep the underlaps of the cord stitch, which have similar orientation and are longer on the back side of the fabrics.

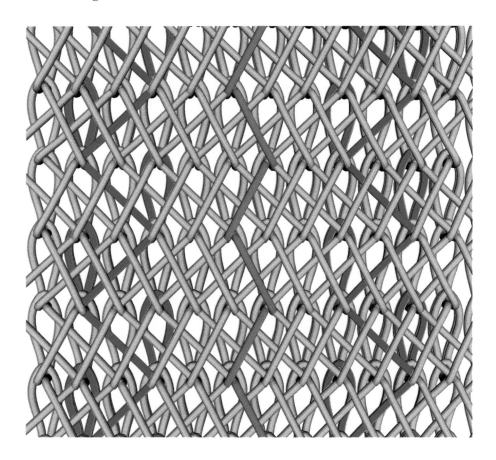

FIGURE 5.11 Tricot-cord with counter-lapping, back side.

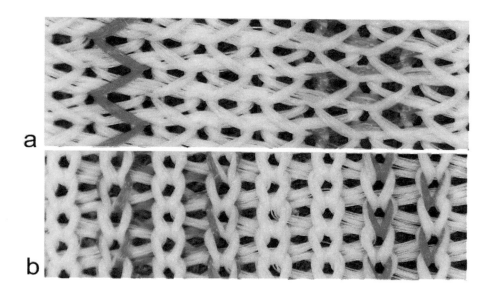

FIGURE 5.12 Tricot-cord with counter-lapping, a) back side with the underlap of the tricot, b)front side. Microscopic image of fabric from [23].

FIGURE 5.13 Tricot-cord with equal lapping, a) back side with the underlap of the tricot, b)front side. Microscopic image of fabric from [23].

5.6   More combinations with laps over one or two courses

In a similar way as the tricot-tricot, cord-tricot and tricot-cord constructions can be built large number of combinations between the different laps for the two guide bars. Figure 5.14a shows satin-tricot and Figure 5.14b velvet-tricot in counter lapping. The properties of these fabrics are similar to the previous explained tricot-tricot and tricot-cord, with reduced elongation. Special colour threading (Figure 5.15) can allow building of vertical areas with completely one and the same colour, which is normally a problem for the combinations of lappings placing loops on two wales.

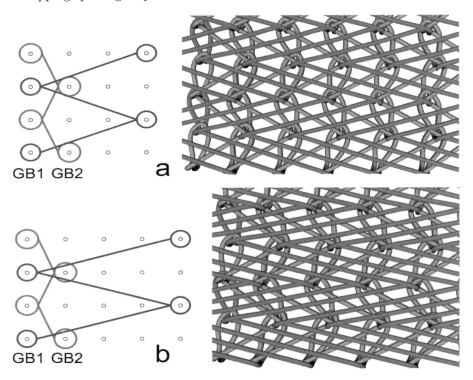

FIGURE 5.14 a)Satin-tricot and b)velvet-tricot in counter-lapping.

The simplest way to get naturally vertical stripes with one colour is the combination between lock-stitch and another lap. The lock stitch has to be performed from the first guide bar, so that its loops are dominating and its

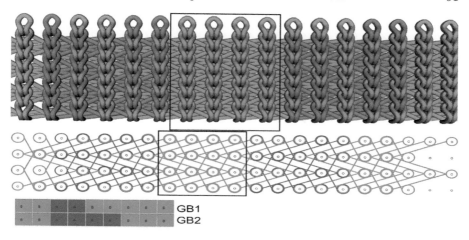

FIGURE 5.15 Satin-tricot counter-lapping with adjusted threading for getting a region with only one colour.

underlaps keep the other underlaps together. This produces very stable structure, which is known as Queen's Cord [18], (Figure 5.16)

A tricot-velvet combination is known as Sharkskin construction (Figure 5.17), where the larger underlaps are trapped inside the structure by the tricot underlaps. This allows rigid control of the flats, prevents any movements of the yarns within the loops and the fabric is extremely stable [18]. The name Sharkskin is being taken from woven cloth with similar characteristics.

The tricot pattern is commonly combined with another pattern, because it is enough to produce stable basic structure with full threaded guide bar. Used as a second guide bar, it remains as hidden stable background. The first guide bar can be used for additional effects like loop-raised structures. For such structures, an equal lapping is used (Figure 5.18c), because the larger instability of this structure has the advantage of larger flexibility here. During shrinkage of the fabrics, the longer underloops appear on the surface (Figure 5.18d) and their single filaments of fibres (Figure 5.18b) can be raised by card-clothed rollers of the loop-raising machine.

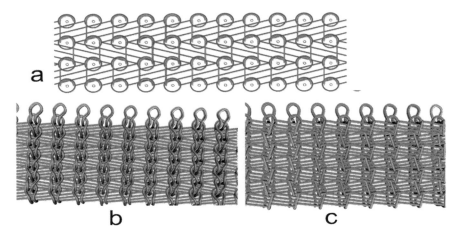

FIGURE 5.16 Lock-stitch -velvet construction, named Queen's cord. a) lapping diagram, b) front side with vertical stripes of wales, c) back side with "pin" stripes.

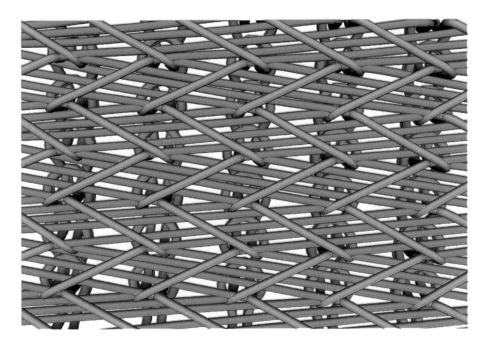

FIGURE 5.17 Sharkskin construction. Tricot underlaps fix stable the long underlaps.

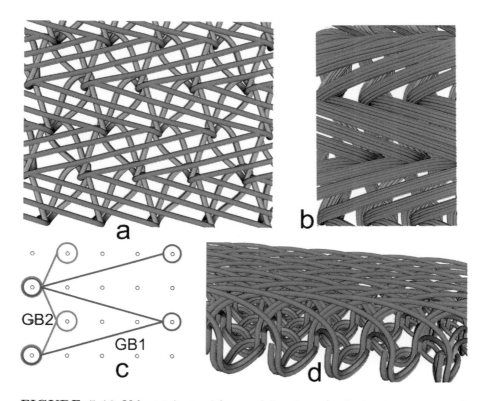

FIGURE 5.18 Velvet-tricot with equal lapping a) idealized 3D view, b) loop-raised filaments of multi-filament yarn of the velvet underlaps c) lapping diagram d) side vew of the the raised underlaps of the velvet yarns.

5.7   Atlas based structures

Atlas lapping on two guide bars with opposite directions (Figure 5.19) are named Double Atlas or Double Nadyke. As in each needle two loops in opposite directions are placed, the loops and the complete wales remain vertical, contrary to the single guide Atlas fabrics. The position, where the guide changes its direction there is a faint horizontal line due to the different loop construction there (Figure 5.20a). Using one yarn of different colour allows typical building for Atlas slight diamond figures, but only the yarn of the first guide becomes visible, as the loops of the second guide and its underlaps bar become hidden behind the first one.

With two yarn colours and suitable threading, different colour effects can be achieved. If both guide bars have the same threading, with a few guides with one colour and the same number with the second colour (Figure 5.21), four colour areas will be available. One area where all loops and underlaps are with colour A, second area, where loops and underlaps are with colour B, and the third and fourth areas will have loops of one colour and visible underlaps of the second colour (Figure 5.22). These four areas can be identified on the back side, too (Figure 5.23). These areas can be manipulated by shifting the threading of the guide bar as visualised in Figure 5.24. In the same way as for the other pattern, the first guide bar determines the loops' colour and the main optical image of the fabrics. If yarn with different colours starts in the opposite position (Figure 5.24a), two diamonds with clear loops and underlaps appear vertically over each other. If the guide bars have identical colour threading (Figure 5.24c), the diamonds are diagonally arranged. If a small overlap of three guides with same colours is arranged (Figure 5.24b) one of the diamonds is reduced in size to the number of the common loops.

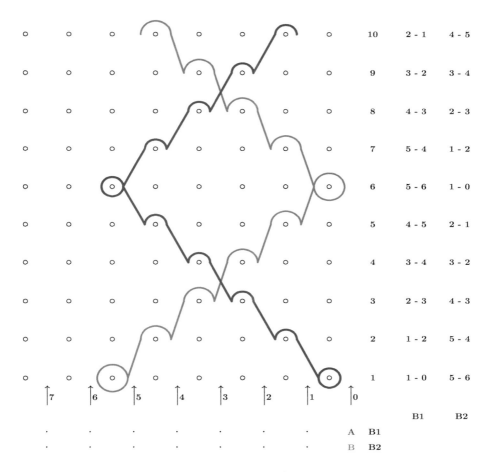

FIGURE 5.19 Double Atlas - lapping movement GB 1: 1-0/1-2/2-3/3-4/4-5/5-6/5-4/4-3/3-2/2-1//; GB 2: 1-0/1-2/2-3/3-4/4-5/5-6/5-4/4-3/3-2/2-1//.

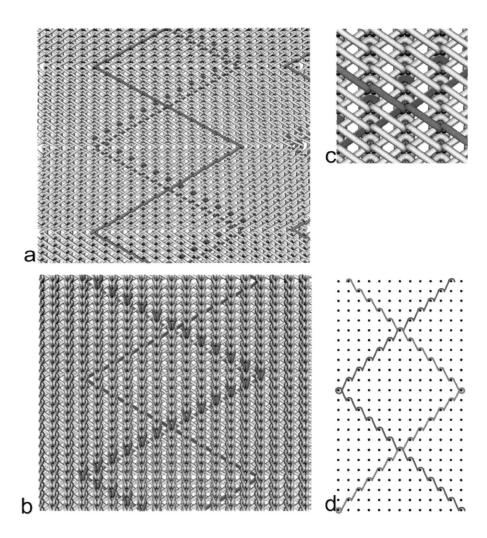

FIGURE 5.20 13 row Double Atlas - a) back side, b) front side, c) view at the crossing area, d) lapping diagram.

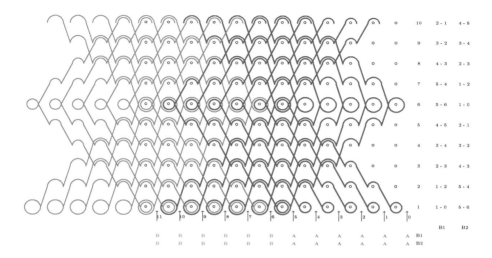

FIGURE 5.21 Double Atlas with identical two colour threading - lapping movement GB 1: 1-0/1-2/2-3/3-4/4-5/5-6/5-4/4-3/3-2/2-1// ; GB 2: 1-0/1-2/2-3/3-4/4-5/5-6/5-4/4-3/3-2/2-1// ; ; Threading: GB 1: A A A A A A B B B B B B ; GB 2: A A A A A A B B B B B B.

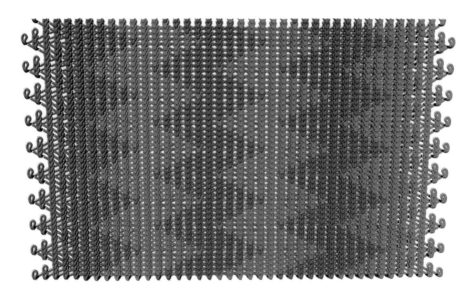

FIGURE 5.22 Double Atlas with identical two colour threading front side - simulation of the pattern of Figure 5.21.

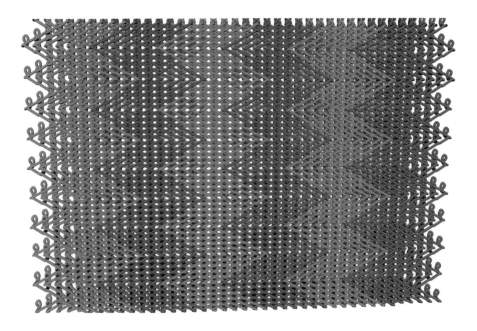

FIGURE 5.23 Double Atlas with identical two colour threading back side - simulation of the pattern of Figure 5.21.

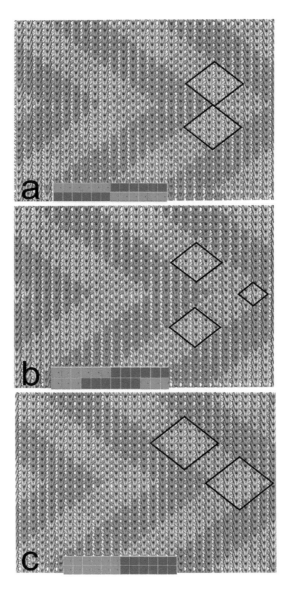

FIGURE 5.24 Double Atlas optics depending on the threading shift of the second guide bar with 6A6B threading a) yarn in opposite arrangement b) three yarns overlap b) parallel threading.

5.8   Tricot Atlas Sequences

Atlas lapping followed by a tricot lapping can be used for a generation of figures with hexagonal colour effects (Figure 5.25). The different areas of the interlacement of the two colours can be recognized as these patterns, too. With some more fantasy and variations of their colours, more effect areas can be created. Nowadays one can lose hours and days, playing with trials of such patterns with the help of modern CAD software, and it is not necessarily all configurations to be described in a book.

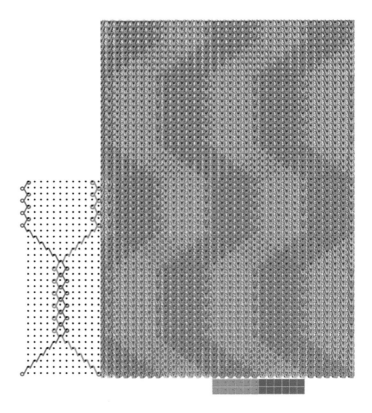

FIGURE 5.25 Atlas-tricot sequence for hexagonal colour pattern. Lapping movement GB 1: 12-13/12-11/11-10/10-9/9-8/8-7/7-6/6-5/6-7/6-5/6-7/6-5/6-7/6-5/6-7/6-5/6-7/6-5/6-7/7-8/8-9/9-10/10-11/11-12/12-13/12-11/12-13/12-11/12-13/12-11/12-13/12-11// ; GB 2: 1-0/1-2/2-3/3-4/4-5/5-6/6-7/7-8/7-6/7-8/7-6/7-8/7-6/7-8/7-6/7-8/7-6/7-8/7-6/6-5/5-4/4-3/3-2/2-1/1-0/1-2/1-0/1-2/1-0/1-2/1-0/1-2// ; Threading: 6A6B parallel for both guide bars.

5.9 Conclusions

Two guide bars with full threading give a very large number of combinations of lappings for different patterns. Some of these are unstable, which means more flexible (mainly with equal lapping), others (usually in counter lapping) lead to very stable structures. The position of the guide bar determines the stability of the structure and the optical and mechanical properties - the yarns of the first guide bar dominate the optics on both fabric sides and keep all remaining yarns on the inside of the fabrics. Using different colour arrangements in the threading creates various effects - such as vertical stripes, hexagons, diamonds and their variations.

6

Two bar fabrics with partial threading

6.1 Introduction

Use of partial threading (Figure 6.1) of the guide bars increases the number of the possible patterns of warp knitted fabrics significantly, but not all combinations with partial threading lead to a stable structure. For any structure where partial threading is used, the information about the threading for the guide bars - which needles are empty and which have yarn - has to be provided in the specification. Partial threading is used for the creation of mesh type structures where smaller or larger openings are available.

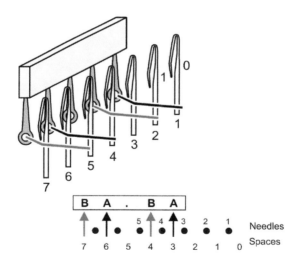

FIGURE 6.1 Partial threading of a guide bar. The third guide does not have a yarn and two yarn colours or types A and B are used. The threading is AB.AB.

6.2   Consistency check of loop existence in each wale

During the design of such structures much attention has to be paid in the consistency of the loops in each wale. If one needle builds a loop during the first row, it has to build loops until the end of the sample. Missing loops in one cycle (Figure 6.2) are (in normal case) not allowed and lead to unknitting of the previously built loops and the structure can be destroyed or exhibit larger openings.

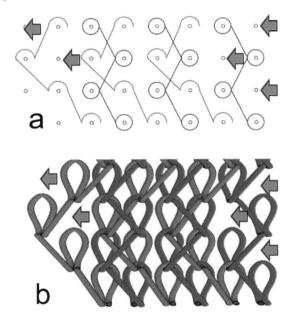

FIGURE 6.2 If the previous loops are not closed in some cycle, they can unknit and increase the size of the opening.

6.3   Different ways to influence the threading

There are different ways to manipulate the pattern for samples with partial threading, as demonstrated in Figure 6.3. The upper combination

(Figure 6.3a) demonstrates double tricot lapping movement for two guide bars, which have half treading - one yarn in one yarn out for both. The two tricot laps in this configuration build vertical loop chains, which are not connected together, but are stable products. Such chains are used in parts of mesh structures. If the one guide bar (in this case the L1) starts instead with "in" with an empty position (".A.A") instead of full ("A.A."), the structure changes significantly - it will consist theoretically of loops, which do not become knitted and practically will not be stable fabrics. The same effect can be achieved, if the threading remains identical for both guide bars, but the number of the positions becomes increased by one (Figure 6.3 c) or its equivalent - if the guide bar receives "offset" of "plus one" as demonstrated in Figure 6.3 d. Theoretically all these three manipulations - of the threading starting point, of the chains and with the offset have identical effects on the structure, but practically some of these require more work as others:

- Threading pattern "shift". Changing the starting position of the threading (Figure 6.3 b) practically requires **new re-threading of the complete guide bar** and is a very time intensive process, especially for machines with larger width.

- Coding change (Figure 6.3 c). The increasing of all numbers in the chain coding today can be done with one command within the modern CAD software such as TexMind Warp Knitting Editor[13] or requires in the worst case, retyping of the numbers on the screen on the machine. For the machines with mechanical chain links, this change requires the building and placement of a new chain, but its change could be faster than re-threading a couple thousand yarns.

- Using offset of the electronic machines (Figure 6.3 d). Using the offset is as well possible in some of the modern electronic control machines. The offset number is entered on the screen or in the software, the pattern remains the same, the real threading remains the same, too, but the guide bar receives a command to "move" always to the given position **plus the offset number**. This solution sounds simple, but is not good practice, because in such cases, the users tend to document the wrong pattern and forget the offset. Generally it has to be used only for quick changes on machine with already prepared yarns for an other sample. Such a solution is possible as well on some of the machines with mechanically patterned devices (chains or cams), where the initial position of the guide bar can be as well adjusted mechanically by extension or shortening of the mechanical bar between the roller and guide bar.

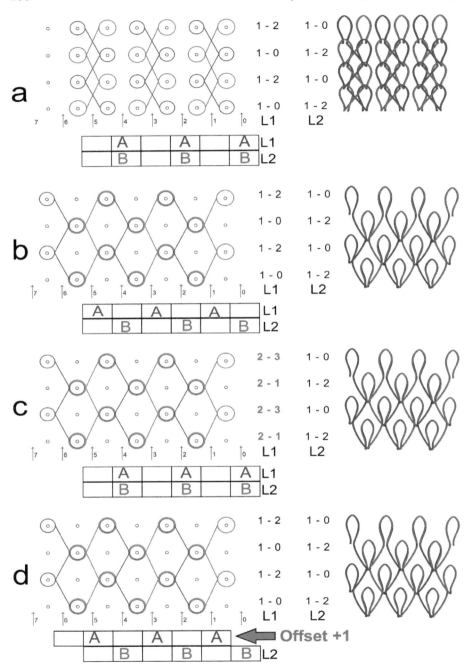

FIGURE 6.3 Chain based on two tricot loops. Lapping movement L1: 1-0/1-2// ; L2: 1-2/1-0//.

6.4 Simple fillet (mesh) structures

6.4.1 Symmetrical structures with half threading 1-in 1-out

Simple fillet structures are produced using two guide bars with partial threading and some of the basic lap in counter-lapping for the guide bars. Figure 6.4 presents such a structure based on a cord stitch. Because of the missing yarns on each second guide, each second loop is not connected to its left or right neighbour loop and openings are built there. The same effect can be reached as well using satin lapping (Figure 6.5). In this case the vertical elements consist of two loops, which, based on the yarn tension, are placed very close together.

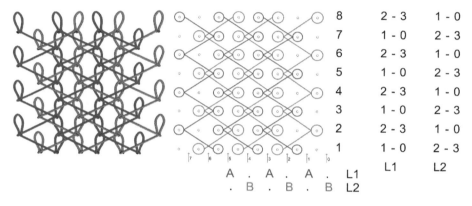

	L1	L2
8	2 - 3	1 - 0
7	1 - 0	2 - 3
6	2 - 3	1 - 0
5	1 - 0	2 - 3
4	2 - 3	1 - 0
3	1 - 0	2 - 3
2	2 - 3	1 - 0
1	1 - 0	2 - 3

A . A . A . L1
. B . B . B L2

FIGURE 6.4 Two guide bars with cord stitch with counter-lapping and half threading build a mesh like structure with small openings. Lapping movement L1: 1-0/2-3//; L2: 2-3/1-01//.

Two atlas laps in opposite directions and partial threading build as well a mesh like structure (Figure 6.6). The three row atlas configurations again with 1-in 1-out threading (Figure 6.7) builds mesh openings with two different sizes - smaller and larger.

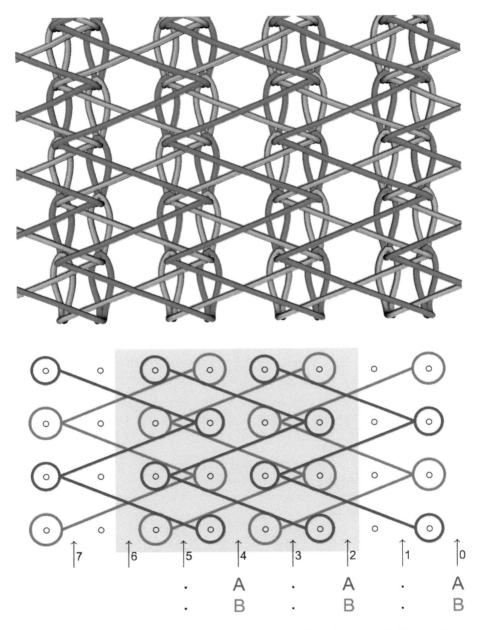

FIGURE 6.5 Mesh with double satin in counter-lapping with half threading GB 1: 1-0/3-4// ; GB 2: 3-4/1-0// ; ; Threading: GB 1: A . A . A . ; GB 2: B . B . B.

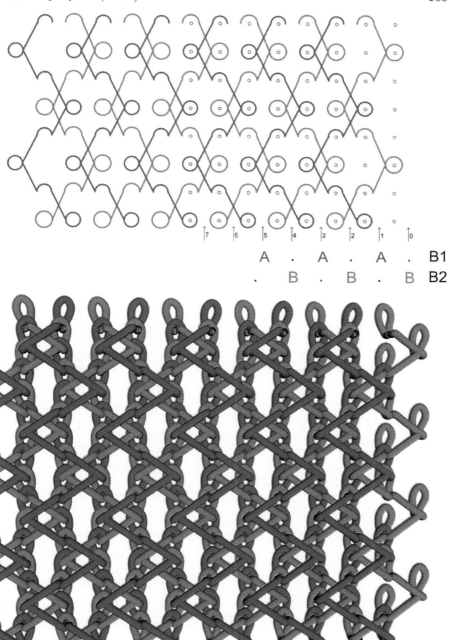

FIGURE 6.6 Mesh with double atlas in counter-lapping with half threading. Lapping movement GB 1: 1-0/1-2/2-3/2-1// ; GB 2: 2-3/2-1/1-0/1-2/// ; ; Threading: GB 1: . A . A . A ; GB 2: B . B . B.

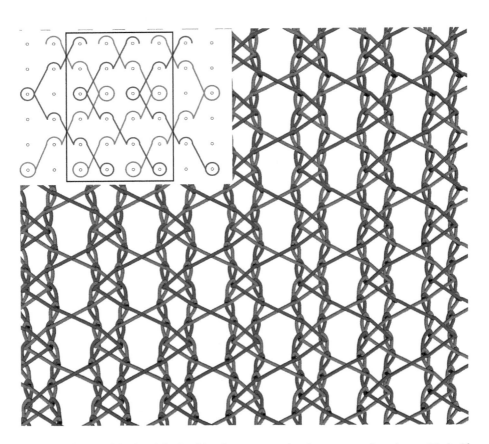

FIGURE 6.7 Mesh with double three-row atlas in counter-lapping with half threading. Lapping movement GB 1: 1-0/1-2/2-3/3-4/3-2/2-1// ; GB 2: 3-4/3-2/2-1/1-0/1-2/2-3//; Threading: GB 1: A.; GB 2: B.

6.4.2 Symmetrical mesh structures with 2-in 2-out threading

The version with back-lapped atlas requires more specialized threading - 2-in and 2-out, arranged so, that the loops are built close together but never on the same needles (Figure 6.8). Such one mesh has a denser appearance because its connecting elements again consist of two loops each (as with satin). A larger view of the fabrics (Figure 6.9) gives a better impression about the character of the fabrics - with specific small openings. Such fabrics are used last time in several areas as face for products, where good air circulation is required for better thermal comfort like shoes, rucksack and other areas.

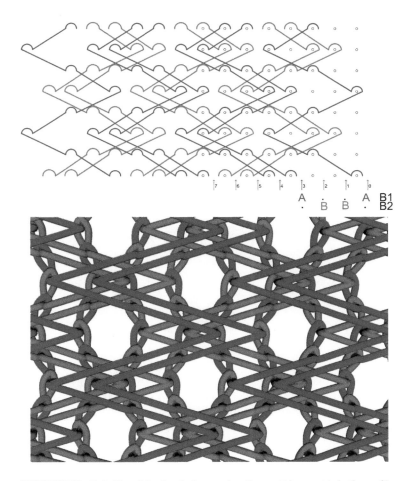

FIGURE 6.8 Double back-lapped atlas with partial threading as pattern and 3D simulation; GB 1: 0-1/2-3/5-4/3-2// ; GB 2: 5-4/3-2/0-1/2-3// ; ; Threading: GB 1: A . . A ; GB 2: . B B.

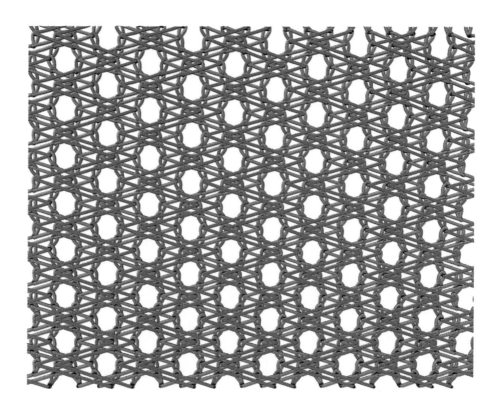

FIGURE 6.9 Double back-lapped atlas with partial threading (Figure 6.8) with more repeats as fabrics' appearance. GB 1: 0-1/2-3/5-4/3-2// ; GB 2: 5-4/3-2/0-1/2-3// ; ; Threading: GB 1: A . . A ; GB 2: . B B.

6.4.3 Asymmetrical structures

Creation of different types of meshes or profiles is possible for a combination of asymmetric laps too. Figure 6.10 visualises such a pattern as a combination of a lock stitch on the first guide bar and a cord stitch on the second guide bar, both in half threading. As the lock stitch is on the first guide bar, its underlaps will go behind the underlaps of the cord stitch and "connected" fabrics will be created (Figure 6.10 b). If the cord stitch is placed on the first guide bar, its underlaps will be as well outside of the lock stitch and there will be no connection between the loops of the both guides. The chains of the guide bar with lock stitch will then be produced and exist independently on the fabrics with the cord stitch from the another guide bar (Figure 6.10 c and d).

Combination of tricot and satin in half threading and counter-lapping leads to a mesh structure, too (Figure 6.11) while changing the combination to equal lapping (Figure 6.12 using the symmetrical pattern of one of the patterns) produces a completely different appearance of the more irregular structure with smaller openings.

One yarn placed with tricot lapping against two yarns in cord stitch with counter-lapping and threading according the Figure 6.13 builds another type of irregular mesh structure with alternating thicker vertical double loop columns and a thin chain of single loops based on the connecting loops of the cord stitches. A combination of tricot with atlas with opposite laps (Figure 6.14) or atlas- atlas (Figure 6.15) can be used for open structures, too. The first one has clearer openings while the second (atlas-atlas) in the current case has a more regular character of the surface due to the larger repeat and the wider areas with connection between the two systems of yarns.

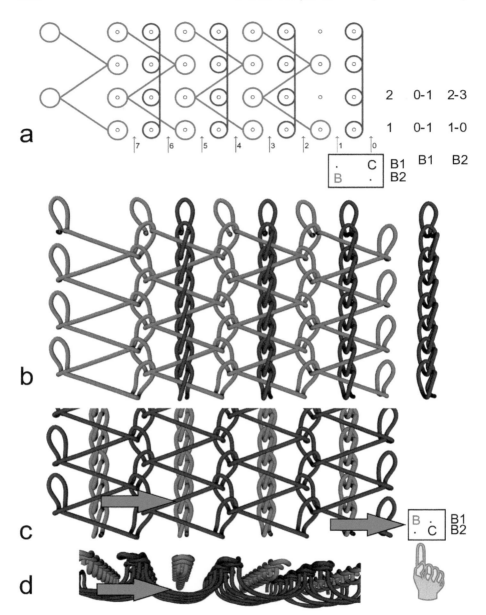

FIGURE 6.10 Asymmetric mesh type structure based on lock stitch and cord stitch a) pattern, b) 3D; Lapping: GB1: 0-1//; GB2: 1-0/2-3//; Threading GB1: C.; GB2: .B; c) If the guide bars exchange their patterns the structure will be not more connected, as the cord stitch underlaps will go behind the lock stitch as seen from top of the simulated structure d).

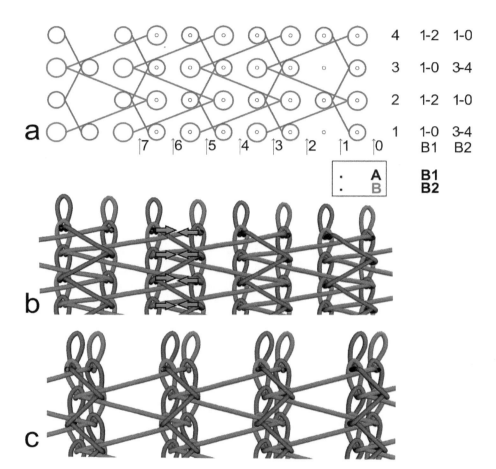

FIGURE 6.11 Tricot-satin with counter-lapping and partial threading, a) pattern and threading lapping movement GB 1: 1-0/1-2/1-0/1-2// ; GB 2: 3-4/1-0/3-4/1-0// ; ; Threading: GB 1:A. ;GB 2: B. ; b) idealized 3D representation. If the yarns have enough tension during the knitting, and they need to have some tension, then the underlaps at this counter-lapping tend to move the complete wales close to each other and the real structure looks like the figure c). The exact position and angle of the loops depends on the tension settings on the machine.

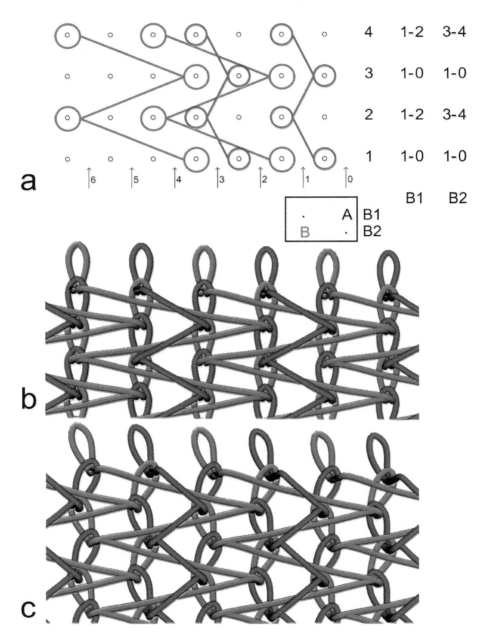

FIGURE 6.12 Tricot-satin from Figure 6.11 but with equal lapping produces fabrics with different appearances a) pattern and threading GB1: 1-0/1-2//; GB2: 1-0/3-4//; Threading: GB1:A.; GB2: .B; b) idealized 3D representation, and c) possible look with slightly moved loops side by side.

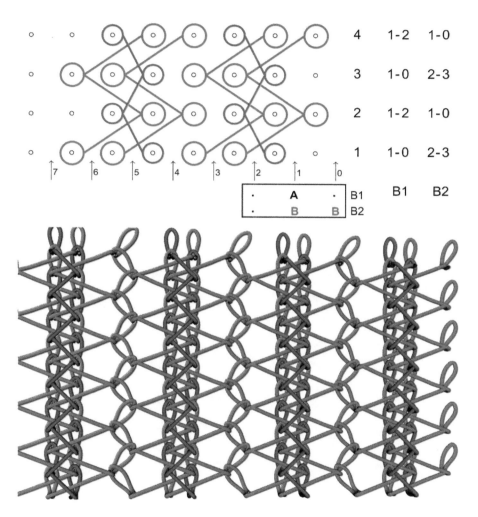

FIGURE 6.13 Tricot - cord in counter-lapping with partial threading. Lapping movement GB1: 1-0/1-2/1-0/1-2//; GB2: 2-3/1-0/2-3/1-0//; Threading: GB1: .A.; GB2: BB.

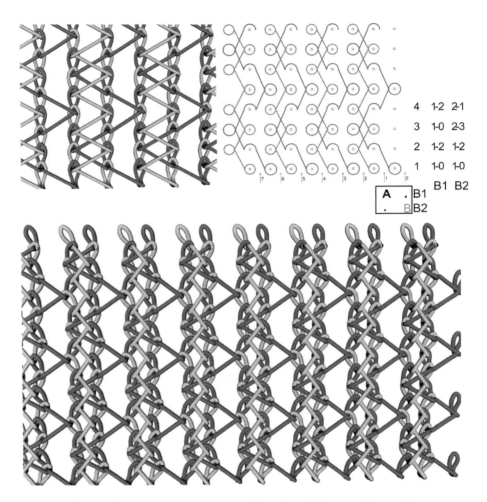

FIGURE 6.14 Tricot - atlas in counter-lapping with partial threading as pattern, topological simulation and larger view with loop orientation correction. Lapping movement GB 1: 1-0/1-2//; GB2: 1-0/1-2/2-3/2-1// ; ; Threading: GB 1: .A; GB 2: B.

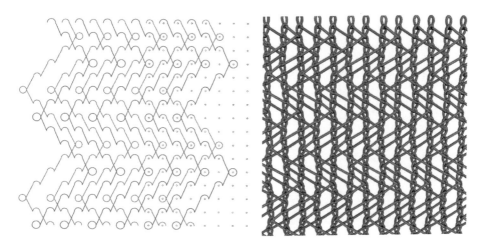

FIGURE 6.15 Atlas - atlas in counter-lapping with partial threading, Lapping movement GB 1: 3-4/4-5/5-6/5-4/4-3/3-2/2-1/2-3// ; GB 2: 4-5/4-3/3-2/2-1/1-0/1-2/2-3/3-4// ; ; Threading: GB 1: .A; GB 2: .B.

6.5 Combined fillet structures

The previously described simple fillet (mesh) structures can be combined together with tricot or lock-stitch based changes for achieving larger openings.

Figure 6.16 presents alternating tricot with cord stitch fillets. It builds larger openings between the tricot chains and small openings in the cord stitch area.

Four rows fillet in alternating tricot and atlas lap is visualised on Figure 6.19. The atlas lap breaks the opening and shifts the vertical chain to the next cell. Using the atlas pattern for connection between the single vertical lines provides smoother transitions for the yarns compared to the connections of the previous two figures (Figure 6.17 and Figure 6.18), where the exact optical appearance in the longer yarn pieces depends on the yarn tension during the knitting.

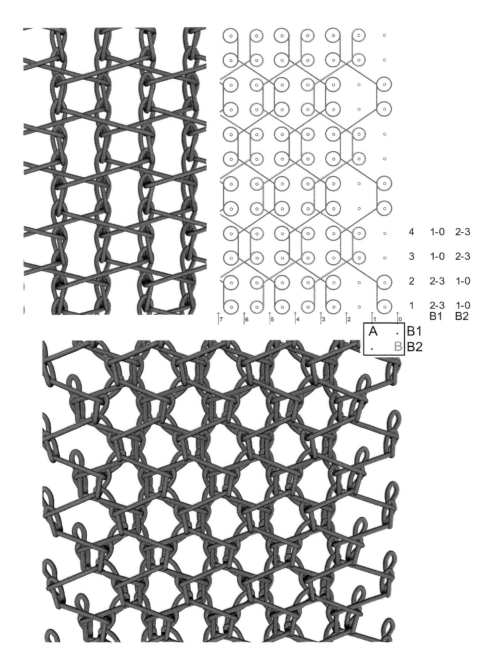

FIGURE 6.16 Lapping movement GB 1: 2-3/2-3/1-0/1-0// ; GB 2: 1-0/1-0/2-3/2-3// ; ; Threading: GB 1: . A ; GB 2: B.

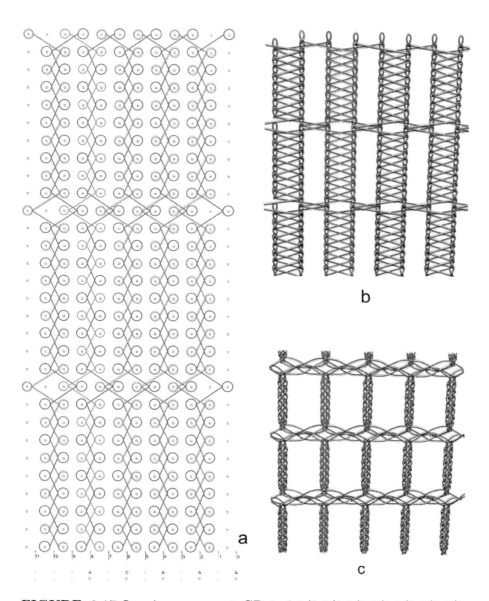

FIGURE 6.17 Lapping movement GB 1: 2-1/2-3/2-1/2-3/2-1/2-3/2-1/2-3/2-1/3-4// ; GB 2: 2-3/2-1/2-3/2-1/2-3/2-1/2-3/2-1/2-3/1-0// ; Threading: GB 1: A . A . A . C . A . ; GB 2: B . B . B . B . B.

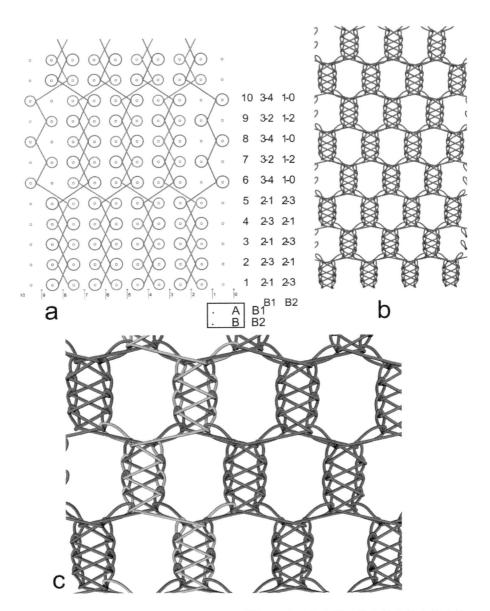

10	3-4	1-0
9	3-2	1-2
8	3-4	1-0
7	3-2	1-2
6	3-4	1-0
5	2-1	2-3
4	2-3	2-1
3	2-1	2-3
2	2-3	2-1
1	2-1	2-3
	B1	B2

a

| . | A | B1 |
| . | B | B2 |

b

c

FIGURE 6.18 Lapping movement GB 1: 2-1/2-3/2-1/2-3/2-1/3-4/3-2/3-4/3-2/3-4//; GB 2: 2-3/2-1/2-3/2-1/2-3/1-0/1-2/1-0/1-2/1-0// ; Threading: GB 1: A. ; GB 2: B.

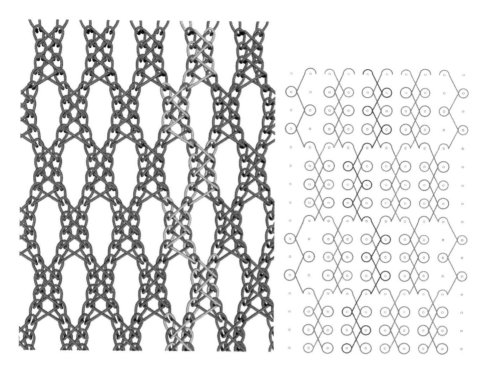

FIGURE 6.19 Lapping movement GB 1: 2-3/2-1/2-3/2-1/1-0/1-2/1-0/1-2// ; GB 2: 2-1/2-3/2-1/2-3/3-4/3-2/3-4/3-2// ; Threading: GB1: A.A.C.A.; GB2: B.

6.6 Conclusions

The part-set or partial threading of the guide bars extends the patterning possibilities of the warp knitted structures significantly and allows the building of structures with openings and larger mesh cells to be created. Important aspects during the development of such structures is the check if the knitting needles (not the empty ones) receive yarns consistently. Mesh type structures normally have larger deformations and different appearances than what is expected from the lapping diagram, but with the help of the modern software this can be simulated quickly.

Part III

Single face structures with more structural elements

7

Laid-in fabrics

7.1 Introduction

The laid-in fabrics are built of at least one ground bar with usually full threading and at least one laid-in guide bar, that does build loops, but only places underlaps. These are very common for crochet-knitting machines for production of elastic and stable meshes, medical and fashion tapes, decoration textiles and as well for the Raschel machines for decorative home textiles. The laid-in yarn is not building loops, it stays in the form of longer floatings - underlaps or like partial weft insertion (which is the other name) between the loops and their underlaps. This configuration is possible (without a falling plate) only if the laid-in guide bars are placed **between** the ground guide bars and the needles (Figure 7.1). The laid-in guide bars often work in partial threading. The samples need different yarn types and colours to become beautiful, but large variations of colours and paths need more guide bars. In order for space for more bars to be provided, pairs of the pattern bars are **nested** together into groups. From the mechanical and machine geometry point of view, one group of two, three or four nested bars takes the same place as one normal guide bar with full threading in the area of knitting needles. During past years, the company of Karl Mayer introduced "string bars" (Figure 7.2b), which need significantly less space in comparison to the older guide bars. The lapping motion (shog) of the guides is performed by strings under tension, which have to be well guided, but which do not require complete moving parts and weight as one solid guide.

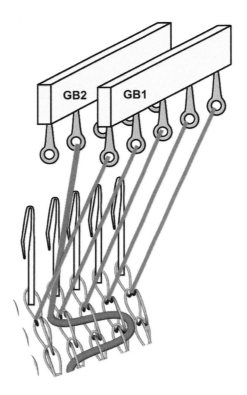

FIGURE 7.1 Principle of building of laid-in fabrics. The laid-in bar places its yarns between the loops and the underlaps of the previous guide bar(s).

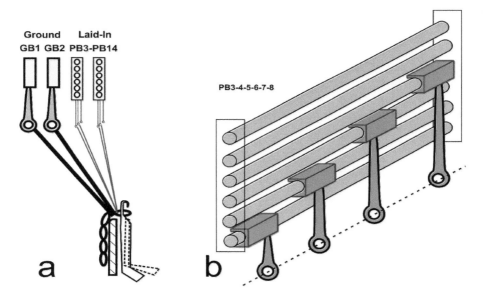

FIGURE 7.2 a) Modern machines can use compact string bars. b) One string bar integrates up to six metal strings. On each string can be clipped one or more guides, which saves a significant amount of space compared to the classical guide bars for laid-in placement.

7.2   Terminological issues

In this chapter the word "laid-in" is used following the terminology of Paling[18]. Other words for the same structural element are *weft*, and especially *partial weft* in order to avoid confusion with *full* weft. Full weft is not discussed in this book because it is connected with insertion of yarns through the complete width of the fabrics using a special magazine weft insertion device [26]. Depending of the number of guides, which place the yarns in the magazine weft insertion device before knitting, several layers under different angles can be placed. This technique is used with high performance fibres like carbon and glass for production of preforms for composites. Another word for the yarn is the *inlay* yarn. The inlay yarns in the braiding have vertical (zero degree) direction, and they can have vertical direction in the warp knitting too, if the guide performs long mislapping. Generally, the laid-in yarn can be oriented in a horizontal (weft) direction and in a vertical (warp or inlay) direction. Here the general case of the type of binding of the yarn in the structure has to be considered; it was decided that the term **laid-in** was to be used.

7.3   Rules for laid-in placement

7.3.1   Basic rules

In the book of Wheatley [27] are described five rules for the determination of the intersections between laid-in yarns and ground structures. These rules provide exact solutions, but are limited to the pillar and tricot stitch in the ground bar. For instance, Rule 1. says: "If the lay-in bar performs equal lapping - moving in the same direction as the ground bar on the **underlap**, then its yarns will be connected to the ground structure by **one less yarn than the needle spaces moved** (Figure 7.3)". Rule 2. : If the laying-in bar moves in the opposite direction to the ground bar during the underlapping, then its yarns will be connected to the ground structure by **one more yarn than the needles spaces moved** (Figure 7.4). The remaining rules are more specific and cover special cases. Paling [18] provides three rules for the intersections, which are always valid, but the number of intersection points are not specified. He says ([18], p.170) "and will be connected to the ground structure by **certain** threads of the front bar".

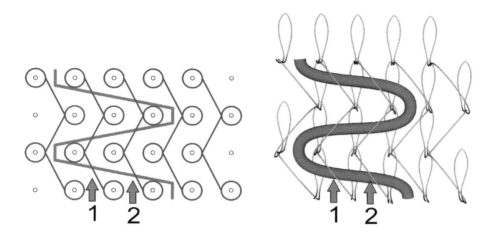

FIGURE 7.3 Ground and laid-in bar with equal lapping. GB 1: 1-0/1-2// ; GB 2: 0-0/3-3//.

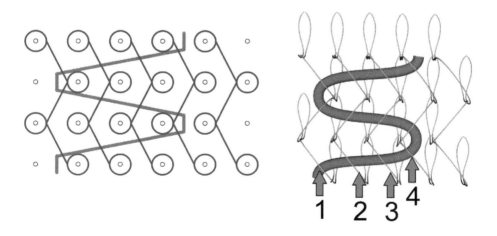

FIGURE 7.4 Ground and laid-in bar with countner-lapping. GB 1: 1-0/1-2// ; GB 2: 3-3/0-0//.

7.3.2 Generalized rule for underlap intersection in laid-in fabrics

Based on the above existing rules in the literature and analysis of several results of simulated structures with the TexMind Warp Knitting Pattern Editor 3D [13], the condition for the integration of the laid-in yarn between the loops and the underlaps of the round and the number of the intersection points can be derived as a single vector equation, describing the relative distance between the guides of the both guide bars (Figure 7.5):

$$\vec{S}_{CP} = \vec{S}_{LaidIn} - \vec{S}_{Ground} \tag{7.1}$$

Here $\vec{S}_{CP}$ is the number of the yarns, which one yarn of the laid-in bar is crossing, $\vec{S}_{LaidIn}$ is vector of the motion of the laid-in bar in number of needle distances and $\vec{S}_{Ground}$ is the vector of the motion of the ground bar, again in needle steps. The $\vec{S}_{CP}$ is the distance, in needle spaces t between the ground and laid-in bar at the end of their underlapping motion. If both bars are moving in the same direction, having "equal underlapping" the distance between these will remain smaller than if they make opposite underlapping. The distance between the guides at the end of the underlapping motion is directly connected to the number of the yarns, which will cross with one underlap. If this number is zero, then the both yarns are not connected and the laid-in yarn is free floating.

The vector sign in equation (7.1) are placed not to confuse the reader; they just have to remind the reader, that the **motion direction** has to be taken into account. Taking into account the direction possibilities, the equation 7.1 becomes translated into two practical equations, which summarize the older rules:

- For **equal underlapping** motion of the two bars the number of the intersecting yarns is equal to the (absolute) **difference** of the underlapping motions of the both bars:

$$S_{CP} = S_{LaidIn} - S_{Ground} \tag{7.2}$$

- For **opposite underlapping** motion the number of the intersecting yarns is equal to the **sum** of the underlapping motions of the both bars:

$$S_{CP} = S_{LaidIn} + S_{Ground} \tag{7.3}$$

If the number of the intersecting yarns S_{CP} is equal to zero, then the laid-in yarn remains unconnected to the ground structure.

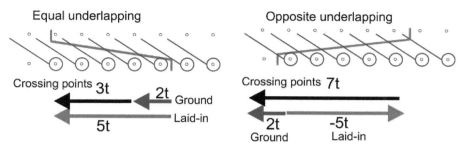

FIGURE 7.5 Figure for derivation of the basic rule for the number of the crossing points and connectivity of the laid-in yarns. Here t is the needle distance. The number of the crossing yarns is equal to the distance between the guides after the motion.

7.3.3 Examples with application of the basic rule

In the case of figure(7.3), the ground bar is making tricot laps with underlaping of one step ($S_{Ground} = 1$) while the laid-in bar is placing yarn through three needle steps ($S_{LaidIn} = 3$). In the case of equal lapping

$$S_{CP} = S_{LaidIn} - S_{Ground} = 3 - 1 = 2 \qquad (7.4)$$

and the laid-in yarn is kept under two underlaps of the ground yarns. In the case of the Figure (7.4) the lapping is opposite,

$$S_{CP} = S_{LaidIn} + S_{Ground} = 3 + 1 = 4 \qquad (7.5)$$

In the same way for the combination between the tricot ground and the equal lapped laid-in yarn with 0-0/1-1// lapping (Figure 7.6a, b), the number of the intersections between the underlaps will be zero, and the red yarns remains **unconnected** into the fabrics:

$$S_{CP} = S_{LaidIn} - S_{Ground} = 1 - 1 = 0 \qquad (7.6)$$

Using counter-lapping (Figure 7.6c, d) leads to two supporting underlaps, which hold the laid-in yarn.

$$S_{CP} = S_{LaidIn} + S_{Ground} = 1 + 1 = 2 \qquad (7.7)$$

Very often as a pillar stitch is used as background, as it has no underlaps $S_{Ground} = 0$; the laid-in yarns are always connected with the number of yarns, equal to the underlapping length S_{LaidIn}, which for the case of Figure 7.7 is $S_{LaidIn} = 3$.

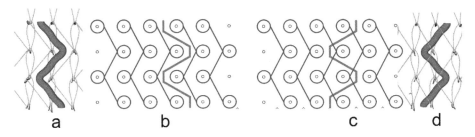

FIGURE 7.6 Figure for derivation of the basic rule for the number of the crossing points and connectivity of the laid-in yarns. Here t is the needle distance. The number of the crossing yarns is equal to the distance between the guides after the motion.

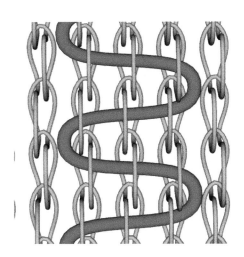

FIGURE 7.7 Pillar stitch is often used a ground for laid-in fabrics.

7.3.4   Laid-in without underlap

All previous cases are assuming that the laid-in guide performs some motion and places its yarns under at least one loop. During all these previous lappings, the laid-in yarns are fixed from one side by the loop legs and the question is mainly the fixation on the underlap side (technical back) to be ensured. If the laid-in guide swings-in and swings-out at the same position and does not perform any underlaps, then the fixation on the back side for its yarns will be missing, as the loop legs are always vertical. Such yarns are not connected (during the period of such motion of the guide bar) in the fabrics (Figure 7.8c); and in order to be kept in the fabric it needs to have loops or weft connection before and after the floating areas.

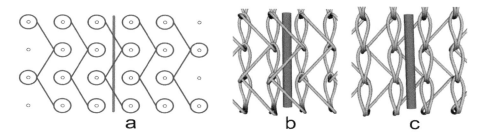

FIGURE 7.8 Vertical not connected laid-in yarns. The crossing of the underlap is not enough to fix the laid-in yarn in the structure, because it is missing a fixing element from the loop side. a) lapping diagram, bc) underlap side - technical back, c) technical front.

7.4 Laid-in as filling

The yarns, placed as laid-in between the loops are largely used for filling different regions and to get solid appearance of it. Lapping motion with larger width behind several needles creates larger (Figure 7.9a) or narrower (Figure 7.9b) solid regions. Lapping in one direction under more needles, than in the other creates diagonal lines (Figure 7.9c). If the yarn has to become hidden, it underlaps only around one needle (Figure 7.9d). Two guides on the same (or another) guide can be used to fill one area together, too (Figure 7.10). In the case of equal lapping (Figure 7.10a), the two sub-areas are not connected and some small gap between these can become visible, which will become larger in the case of counter-lapping (Figure 7.10b). The tension force in the laid-in yarn during the knitting can cause side movement of the loops at the turning points and the intensity of the movement depends on the tension of the yarn, which is related to pretension and the run-in values. For denser filling of the regions multiple laid-in yarns are used (Figure 7.10c). Counter-lapping with less common filling without gaps can be obtained by laid-in yarns which turn their direction around the same loops (Figure 7.10d).

All these types of lapping are demonstrated on several samples in this chapter and in the chapter about jacquard fabrics, Chapter 9.

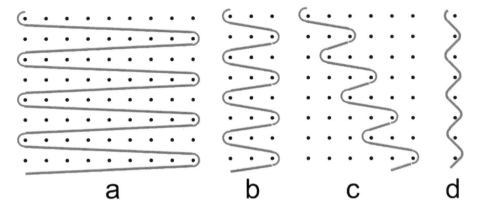

FIGURE 7.9 The laid-in yarns can fill regions with different sizes and ge-ometry or can remain hidden a) larger region, b) connecting three wales, c) producing angled thicker line, d) hidden between the loops.

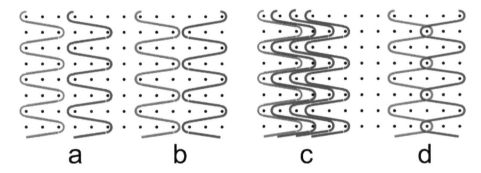

FIGURE 7.10 Types of filling with laid-in pattern - a) not connected areas with equal lapping b) not connected areas with counter lapping c) filling area covered by multiple yarns d) connected areas with counter-lapping.

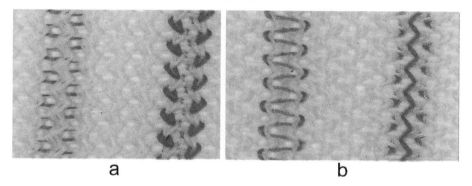

a   b

FIGURE 7.11 Laid-in yarns as filling a) loop side b) underlap side, lapping: loops closed tricot 1-0/1-2//, laid-in 0-0/3-3//, sample 48 from [23].

Figure 7.11 visualizes both sides of plain tricot structure filled with laid-in yarns for stabilisation or for preparation for napping in order to produce rough surface. The samples where the filling is used for decoration follow in the next sections.

7.5   Half-mesh structure - alternating loop and laid-in

An interesting configuration appears if a guide bar (with full threading) alternates placing loops and laid-in (Figure 7.12a). The yarn builds one loop and after that slides around the underlaps until some fixed position, the initial geometrical configuration is simulated in Figure 7.12b. The structure consists of connected (interlooped) loops and interlaced slings Figure 7.12c and Figure 7.12d and has the German name "Schlingenstoff", which in translation means sling based material.

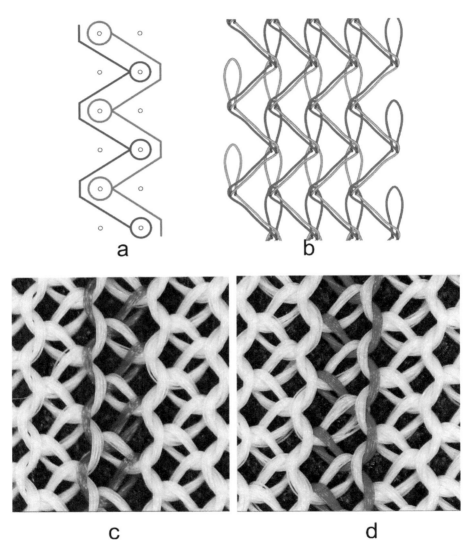

FIGURE 7.12 Half mesh structure, alternating loop and laid-in in each guide bar a) lapping diagram, b) idealized simulation c) and d) photograph of sample 53 of [23].

7.6 Laid-in fabrics as rectangular net

7.6.1 Ground structure

Except as filling area, the laid-in is largely used as a connecting element between the loops in both meshes with rectangular or hexagonal or other more complex geometry. Such products can be used as a ground for laces, curtains and for a large number of other applications. The laid-in yarns limit the elongation of the fabrics in horizontal direction similar to the weft yarn in woven fabrics. If additional laid-in yarns are inserted vertically (Figure 7.10d), almost inextensible in both direction fabrics with similar to the woven structures can be produced. With possible lapping motions for the two guide bars curtain mesh is presented in Figure 7.13a. For full threading of the both bars, a dense structure can be produced (Figure 7.14). With half threading of the laid-in guide bar and lapping movement as presented in Figure 7.13b a typical marquisette mesh can be produced, and this is very commonly used as background structure for laces. Each laid-in yarn, which goes through several wales horizontally stabilizes the position of the vertical chains and gives very good stability of the structure (Figure 7.15). Such meshes are used as net structures against insects or animals on windows, in agriculture and other areas. The structure looks like a rare woven structure, but because of the interconnection of the laid-in yarn between the loops has a significantly higher seep slippage resistance than woven fabrics.

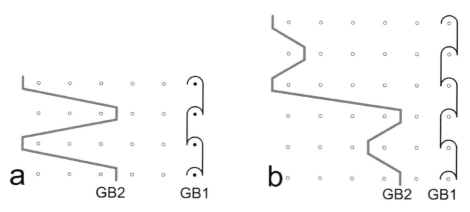

FIGURE 7.13 Pattern for curtain mesh with laid-in yarns.

In order to get more equilibrium of the forces in laid-in fabrics and keep the mesh openings stable, two guide bars in counter-lapping with laid-in yarns have to be used (Figure 7.16). In this case the lock stitches (Figure 7.16a) are connected at regular distances with the laid-in yarns (Figure 7.16b) and (Figure 7.16c), which in the remaining time remain hidden around the loops

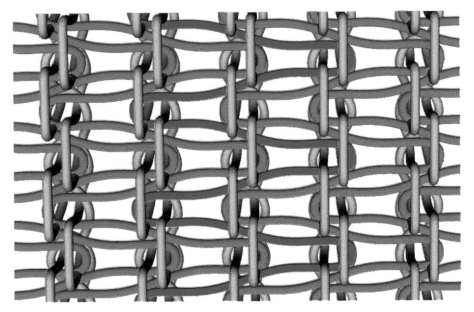

FIGURE 7.14 Curtain mesh at full threading, lapping movement of Figure 7.13a.

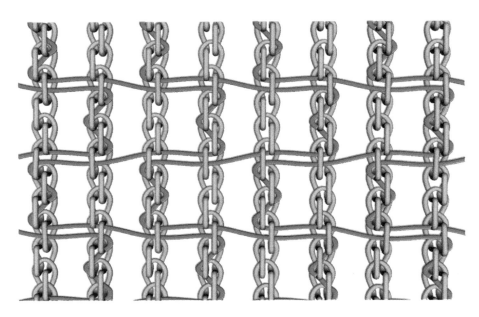

FIGURE 7.15 Marquisette curtain mesh, lapping movement of Figure 7.13b.

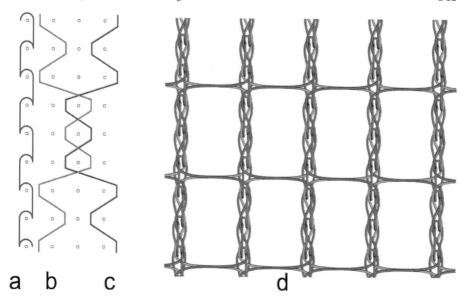

a   b   c   d

FIGURE 7.16 Three bar marquiesette net as rectangular ground mesh. a) lock stitch b) and c) connecting laid-in yarns.

and limits the elongation of the fabrics in a vertical direction. The width of the rectangular cells (Figure 7.16d) can be adjusted by using partial threading for the lock-stitches (using another one or skipping two or more needles), as in this case the lapping movement of the laid-in yarns has to be adjusted to reach the next wale. The height of the cells can be adjusted using more loops in the chains. Figure 7.17 presents another variant of this pattern, where one of the yarns after leaving the vertical chain, passes through the next chain and then connects with the third one. This construction has a higher strength and stability in the horizontal direction without the use of more yarns in the vertical direction.

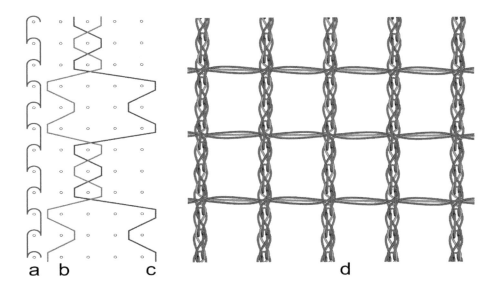

FIGURE 7.17 Three bar marquisette net as rectangular ground mesh. a) lock stitch b) and c) connecting laid-in yarns while the laid-in yarns of the guide bar c) connects two wales.

7.6.2 Lace patterning

Figure 7.18 presents a zoomed example for preparation of the pattern for lace, based on a two-bar background, based on lock-stitch and laid-in connections. The background is drawn less visible, and the accent is given to the decorating laid-in yarns which builds some figures. The simulation of the complete structure including the background structure and the decorating laid-in yarns was generated using the TexMind Warp Knitting Editor 3D and is given in Figure 7.19. Normally for lace structures a photo realistic simulation is required, which does not have to be in 3D but has to present the character of the surface of the structure more and the specifics of the yarn material. This can be done automatically in specialized lace development software or using additional graphics tools, where the light and the optical properties of the yarns have to be specified in detail.

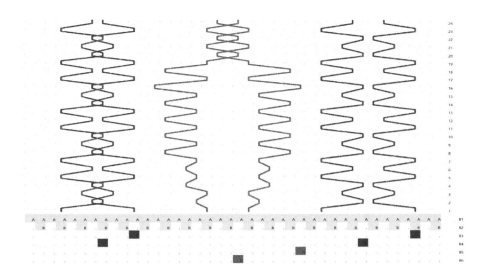

FIGURE 7.18 Pattern for lace structure based on marquisette curtain mesh, developed with TexMind Warp Knitting Editor 3D.

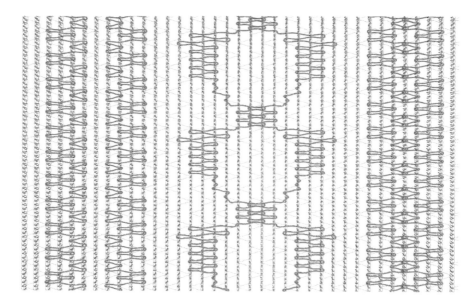

FIGURE 7.19 Raschel lace 3D simulation, lapping movement of Figure 7.18.

7.6.3   Lace examples with seventeen and nineteen bars

The following figures demonstrate a sample, produced on E24 machine with nineteen guide bars. The wider regions of the pattern are filled from two or more guides (Figure 7.20) with some intersection of the paths in order to ensure the connectivity of the regions. In this way the path length and the resulting dynamical loads over the guides and pattern drive can be reduced. Additionally, well selected boundaries of the single regions can be used to intensify the optical appearance with additional curves. The filling degree can be varied by using yarns with different fineness, as demonstrated in Figure 7.21. In this figure the yarn for the upper filling area is twice as fine (150dtex f96x1) than the yarn for the bottom area (150dtex f96**x2**). As a result, the bottom area appears denser and less transparent. In the areas between the filed regions the chains change their position in order to get to a relaxed state with minimal energy. Depending on the run-in settings, the yarn bending stiffness and the pattern these reorientations can be influenced. In Figure 7.21 the laid-in yarn is thicker and probably placed with a little bit of higher tension than the loop building yarns, which causes side inclination of the chains to the direction of the laid-in yarns at their turning points. Effective usage of this technique can be used as a part of the design of the complete sample, shown in Figure 7.22. A similar sample of the same material is shown in Figure 7.23, where some of the vertical chains are deflected alternating to the left and right side to build hexagonal elements (Figure 7.24).

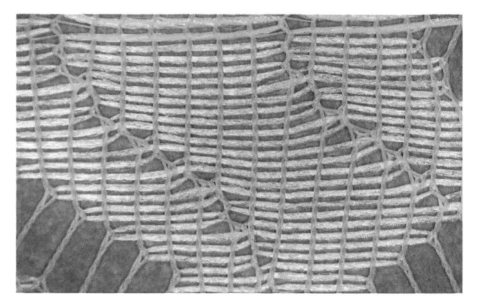

FIGURE 7.20 Larger area filled by three guide bars. The common edge is arranged as curve, which is parallel to the outer counter. Part of Figure 7.22.

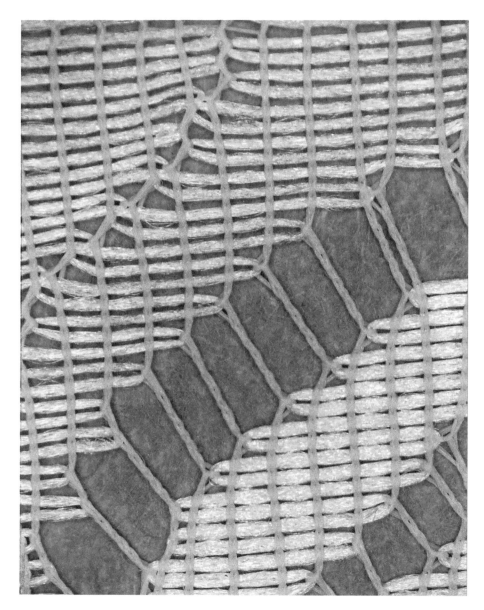

FIGURE 7.21 Yarns with different fineness for variation of the filling rate in the upper and bottom regions. Part of Figure 7.22.

FIGURE 7.22 Sample 12771 of the Karl Mayer company, produced with 19 guide bars on E24 machine.

FIGURE 7.23 Sample 12773 of the Karl Mayer company, produced with 19 guide bars and selective connection of the ground chains for building hexagonal effects.

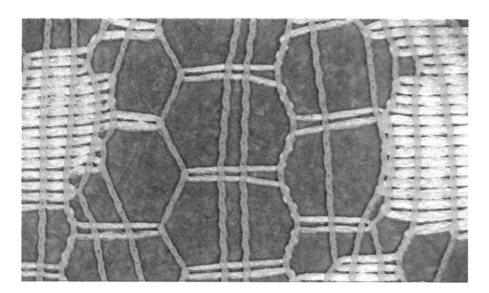

FIGURE 7.24 Deflecting the chain with laid-in connections. Magnified part of the sample from Figure 7.23.

7.6.4 Lace example with triangular effects in the connection areas

The creative use of laid-in yarns allows not only filling areas and side movement of the chains, it allows production of larger figures as well based on thinner and thicker lines. Figure 7.25 demonstrates in magnification the nice "circle" from the Figure 7.28. It is created by a combination of thin lines based on longer floats of the laid-in yarns and thick lines, produced step-wise returning forward or backward moving laid-in yarn (0-0/2-2/1-1/3-3/2-2/4-4/ etc.). The flowers are filled in the standard way (Figure 7.26) and an interesting, triangle-like effect is achieved for the background (Figure 7.27). The triangles have as basis a linear area with very large laid-in placements (0-0/8-8/), built of the yarns of three or more guides. After this area the lapping width is reduced and finally only one laid-in yarn connects two vertical lock-stitch chains, which means that the laid-in bar has a half threading. Because the two chains are fixed well in the basis line through the friction with the loops and the yarn tension of the laid-in yarns is kept low enough, the two chains approach each other slowly and finally appear as a single chain. The laid-in yarn goes left to the next top after reaching the top of the triangle, and because the laid-in yarn from the neighbour chain arrives there, the two chains remained fixed vertically. After that there is no more laid-in yarn, which keeps the two chains close together and they start to move away again. Optically the chains remain less visible and the customer receives the impression of the triangles, flowers and lace type circles (Figure 7.28).

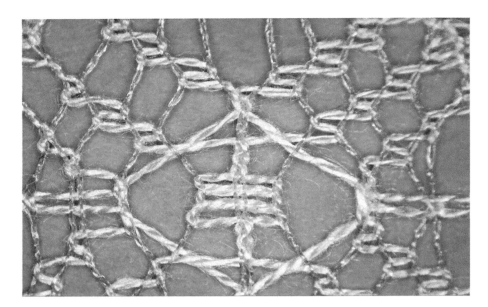

FIGURE 7.25 Figure, created by combination of straight laid-in yarns and lines, created by laid-in lapping. Magnification of Figure 7.28.

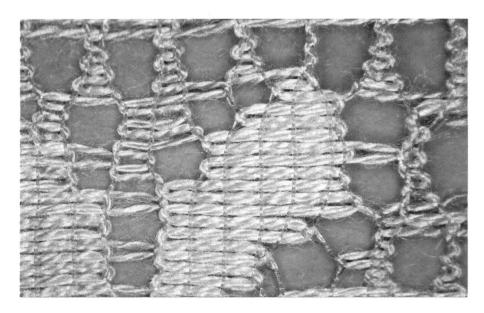

FIGURE 7.26 Standard filling with laid-in yarns, magnification of Figure 7.28.

FIGURE 7.27 Ground effect with triangles, magnification of Figure 7.28.

FIGURE 7.28 Sample 32106/512 of the Karl Mayer company, with triangles background and two types of flowers. 22 Guide bar used, on machine E18. Ground 33dtex polyamide, pattern polyester Nm50/2 staple fibre yarn.

7.6.5　Edges intensification

The laid-in filling in the laces creates edges in the form of waves, because of the turning points of the usually thicker laid-in yarns. In order to get smoother and more intensive appearance of the edge, the separated yarns places, again as laid-in to some of the borders (Figure 7.29). The microscopic image of this sample (Figure 7.30) visualizes the edge yarn, textured polyamide 100dtex f30**x6**, which is used as a boundary for the area, filled with the four times finer 156dtex f34x1, again textured polyamide. On the right part of the figure a third, semitransparent region can be recognized with laid-in of 78dtex f17x1 textured polyamide. Together with the openings, where the chains are not connected, on this sample three filling regions and one edge yarn are used to create the complete design.

FIGURE 7.29 Sample 12610 of the Karl Mayer company, with regions of three different fillings and figure edges made by additional yarns. Total of 32 guides are used for this sample.

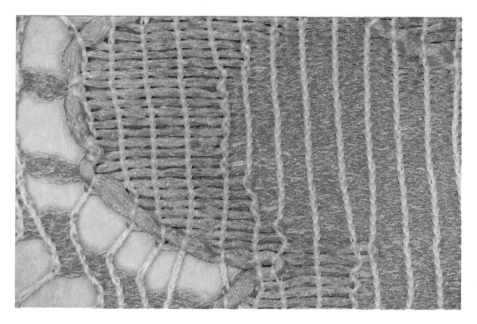

FIGURE 7.30 Magnified part of the sample from Figure 7.29.

7.7   Integrating the decorative laid-in yarns in the ground effects

The previous grounds, based only on lock stitches are simple for production and understanding, but have some difficulties in the patterning because the chains do not remain stable during production. This instability can be used effectively for creating some effects, but its control is not simple. Such samples can be found from the earlier time of production of knitted laces, where the machines had just few guide bars [18]. In the case, where the background has to remain more stable, one or two **additional** guides are used to create the basic mesh. Figure 7.31 presents a magnified small area of the sample of Figure 7.32, where the vertical chains with lock-stitch are connected at every fourth loop with the mono-filament laid-in yarn. The patterning guides fill the design areas in the common way, but in the transitions between the figures builds horizontal lines with small crosses as an additional effect.

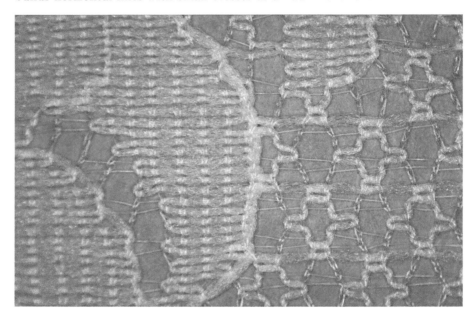

FIGURE 7.31 Two bar ground, reinforced by figures of the patterning laid-in. Magnification of the sample from Figure 7.32.

FIGURE 7.32 Lace over two bar ground, with additional figures based on the laid-in yarn. Sample DO539 of the Karl Mayer company, produced with 32 guide bars on 24E(=48RE) machine.

7.7.1   Connection of the laid-in yarns as figures only

The yarns for the laid-ins are usually significantly thicker and coarser, in order to create the effect of visible regions. After the figure is filled, the yarn has to be lead to the next area, where it will build again a visible figure, but not in all designs are thicker effects as in the previous example from Figure 7.32 welcome. There are known machines with cutting devices, which can automatically cut the laid-in yarn after it has finished the design region and it starts knitting again at the next region. In the classical way the laid-in yarns have to be integrated somehow in the structure, and this is done by well prepared designers' paths, so that they do not disturb the figures. Figure 7.35 demonstrates such a sample with the text of ITMA from the past year 1987. The connection of the thicker area with the ground is visible in Figure 7.34. The figure uses three different yarns for the effects - finer and coarser glittering yarns (Figure 7.33) and one fine, not glittering yarn (Figure 7.36) to create the visual effects.

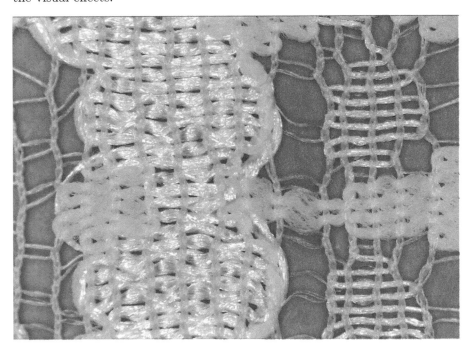

FIGURE 7.33 Use of fine and course yarns for different filling effect, microscopic view of sample Figure 7.35.

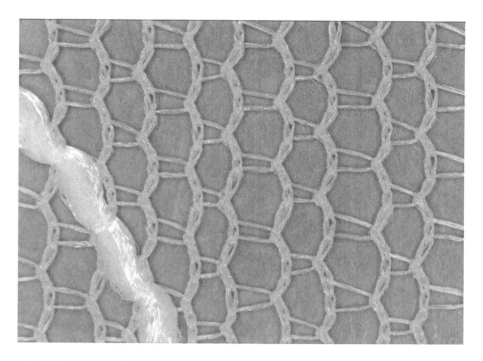

FIGURE 7.34 Ground and piece of the effect yarn connected in it. Microscopic view of sample Figure 7.35.

FIGURE 7.35 Integration of the laid-in yarns as part of the figure design. Sample S-42035 of the Karl Mayer company. Produced with 42 guides and SU patterning drive. Design of M. Göbel-Becker, 1987.

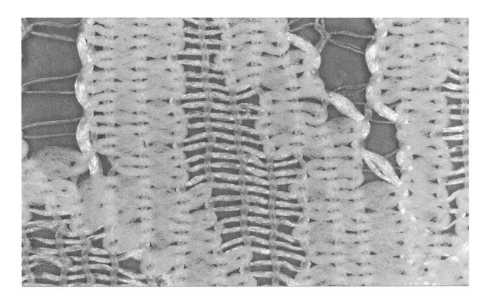

FIGURE 7.36 Combination of yarns for visual effects - 110dtex f51x4 and 110 f51x1 textured and glittering; and 110 dtex f51x2 without glittering surface. Microscopic view of sample Figure 7.35.

7.8 Octagonal mesh with lock stitches

7.8.1 Lace example

Laid-in yarns can connect the lock-stitch chains at **different** steps, for instance, after three loops to make a connection to the right chain, then come after two loops back, then connect to the left chain and then come back again to its own lock-stitch chain. Due to the side inclination of the chain, the resulting structure has eight edges (octagons) (Figure 7.37). The finer laid-in yarn produces half-transparent fillings (Figure 7.38) and the double coarser fills the gaps very well in the structure giving the solid structure a in transparent appearance. Often the laces are cut at not at the rectangular edge, as shown in the sample Figure7.39).

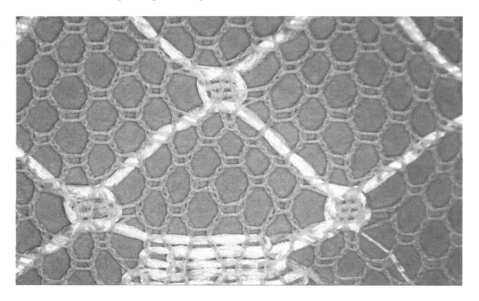

FIGURE 7.37 Lock stitch based ground with two laid-in yarns, which build octagonal cells as ground. Microscopic view of sample 7.39.

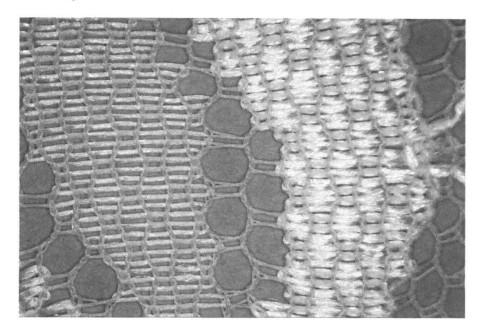

FIGURE 7.38 Fine laid-in yarn builds half transparent area and the coarser yarn fills the gaps completely. Microscopic view of sample 7.39.

FIGURE 7.39 Sample S560055 of the Karl Mayer company, with octagonal ground, two different densities and fancy edges.

7.9   Hexagonal meshes

7.9.1   Pattern

Similar to hexagonal meshes, based on two guide bars, which places yarns for loops, a hexagonal meshes can be built as well with one guide bar with loops and one guide bar with laid-in yarns. For these samples the loops and the laid-in-yarns "go everywhere together". The loop makes connections to the next loop or the next chain and the laid-in yarn stabilizes the structure. Example of lapping movement for pattern with three- and five-course net holes are given in Figure 7.40. If the main element of the chains, which consists of two loops (Figure 7.40c) is repeated once more, the size of the element gets longer. The tricot stitch after the chain is applied in order to jump to the next wale. The resulting structure will have topologically vertical chains which are connected every few courses (Figure 7.41a), but after the relaxation of the yarn it forces every another chain to take a position between its upper and bottom two, moving to one side (Figure 7.41b). The resulting hexagonal mesh is often used as a background for different structures. During the pattern design it has to be taken into account the original position of the loops in order to plane the proper needle gap. All points of the hexagonal cell denote the positions of the needles, which means that the right hand side of the grey filled hexagons all use the same needles. If one guide turns its direction within the grey cell, it always has to have lapping "2-2" there. For simplification of the drawing, all the points of one needle are connected and the complete hexagonal cells are not drawn (Figure 7.42). The five courses mesh has correspondingly five dots per line. Important for the pattern design is that the relation between the width and the length of the cells has to correspond to their size after the fixation and finishing of the ready product, so that the designers have the correct aspect ration in the representation of the figures. Microscopic image of a two course hexagonal mesh is given in figure 7.43. The appearance and the exact size of the mesh depends on the yarn tension and the run in during production and the knowledge of the pattern only is not enough in order to get the correct geometry. Small changes of the relaxation behaviour of the filaments due to no constant spinning conditions can lead to visibly different mesh sizes.

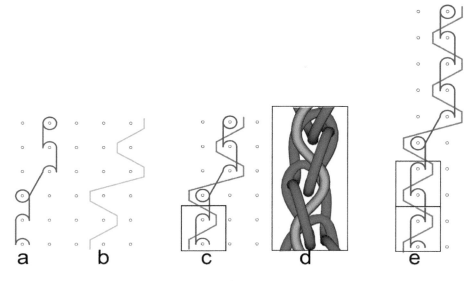

FIGURE 7.40 Lapping movements for three and five course net holes ground structure. Lapping diagram for a) the loops, b) laid-in guide c) both together with marked repeating cell, d) the cell as 3D simulation, 3) five course hole.

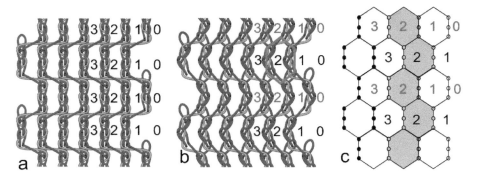

FIGURE 7.41 Three course net hole structure a) topologically correct simulation - it presents the product as created on the machine b) *slightly* deformed to demonstrate the building of the hexagon cells c) drawing for creation of the pattern with identification of the gap number in the cells.

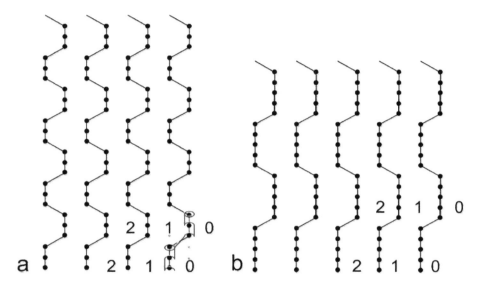

FIGURE 7.42 Mesh for creating the pattern for a)three course hole hexagonal background b)five course hole hexagonal background.

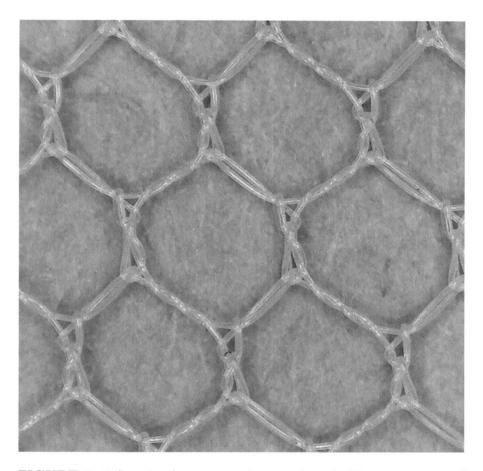

FIGURE 7.43 Sample of two course hexagonal mesh. Microscopic view of sample 12188 of the Karl Mayer company.

7.9.2 Example structure

Figure 7.44 demonstrates a magnified piece of the sample of Figure 7.45 with hexagonal mesh, produced on a machine with gauge E24 according the specification with 22dtex f1 polymid, actually not visible on the microscopic view or not correctly denoted in the documentation; 400dtex polyamide for the loops and 100dtex f32x2 textured polyamide for the laid-in filling. The sample required 32 guide bars and the filling area of each one had to be created carefully with thinking and finding ideas where and how the yarns have to be integrated in the sample after the figure ends.

FIGURE 7.44 Ground and decoration elements of the sample Figure 7.45.

FIGURE 7.45 Sample with hexagonal ground mesh and laid-in yarn decorations.

7.9.3 Elastic ground

Modern humans use a large number of elastic products, which are produced on warp knitting machines, too. There several ideas and possibilities for integration of hyperelastic material (Elastan) in the structure and this is done usually as laid-in yarn. The laid-in yarns, which leads to stabilisation of the previous described structures can be replaced by hyperelastic material. Such material requires positive feeding, so that there always is the same amount of it integrated between the loops under tension. The higher the initial deformation of the elastic yarn during the integration process is, the higher the elasticity of the final product. Lapping movement for such for guides for elastic mesh is visualized in Figure 7.46, based on [18]. The two loop building guides are making one loop and then tricot lapping to the next one and then comes back to their initial course (Figure 7.46a and b). The laid-in yarns are integrated around each chain using a simple 0-0/1-1// motion (Figure 7.46 c and d). The chains' straight vertical loops (Figure 7.46e) come after the production is deformed at the connection places into a zig-zag form, which builds a mesh like structure with openings (Figure 7.46f).

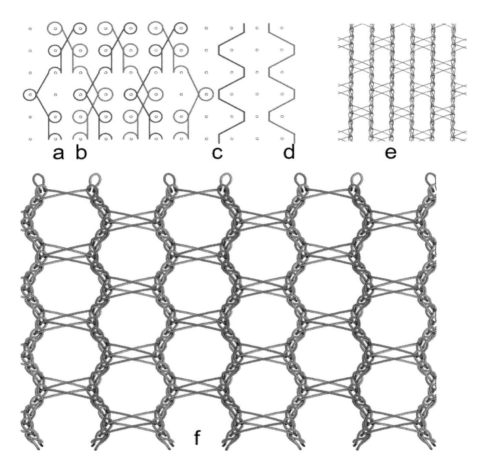

FIGURE 7.46 Four bar elastic ground for mesh fabrics. a) and b) build loops, c) and d) laid-in usually with hyperelastic yarns with positive feeding. e) simulation.

7.10   Conclusions

This chapter demonstrated main, but not all possibilities for patterning with laid-in elements. They are very important elements in warp knitted structures and are used both as connecting elements and as patterning elements. The less elastic laid-in yarns reduce the elongation of the structures and can make them stable in both wales and course direction, so that they have the same stability as woven structures. Hyperelastic yarns are placed in the structures as laid-in elements in order to make these elastic for close to the body clothing.

8

Additional patterning possibilities

8.1 Introduction

This chapter addresses some techniques, which are not described in the remaining part of the book. It covers the fall plate fabrics, where stitching-like effects are created on the yarn surface. The older cut presser technique is mentioned. Finally the principle of creating of samples with pleating using an electronic beam control is discussed.

8.2 Fall plate fabrics

One often used technique for making stitching-like effects on the warp knitting machine is the use of a fall plate. The yarns are connected between the loops and underlaps at the turning points of the motion and built so named "tucks". The plate is a thin plate or blade, placed between the ground building bars and the tuck guide bars, which falls down after the overlapping step and slides the yarns out of the needle hock, so that they cannot build loop. The yarn pieces then initially plate over the previous loop and after the knitting cycle remains connected in the structure. The complete process is explained in detail in almost all books on warp knitting technology like [24], [18],[26],[5], etc. Detailed variants of the different types of connections between the ground structure connecting are given in the book [26], actually in German language. Normally the build tucks are used for decorative applications, and it makes sense that string bars are to be used in the front of the fall plate, Figure 8.1. It is important that the machine has ground building guide bars which creates the background structure, where the tucks are connected. There are different types of connections of the tucks within the structure depending on the lapping motion and the type of the loops in the basic structure. In all kinds of connections, it is important that the tucks are connected with the loops only on the turning points (Figure 8.2b and c) and that they are **floating** between these turning points in the back of the structure (which will be used as front in this case). The laid-in yarns, on the contrary, are connected during their complete path between the loops and underlaps (Figure 8.2a). As the tucks do not build loops, a thicker, cheaper or effect-yarns can be used in similar way as for laid-in yarns. Because of the floats between the turning points, the optics are similar to a stitched structure. If the lapping of the guide bar is like an *open* loop, the tuck is connected similar to the laid-in connection at the turning points but within the head of the loop of the *previous* cycle (Figure 8.2b). If the tuck is placed as a closed loop, it takes a more complicated form, which idealized, or relaxed geometry is presented in Figure 8.2c. In the lapping diagrams the tucks are drawn usually as a single loop so, that the person who reads this documentation is able to reproduce the lapping motion based on the drawing.

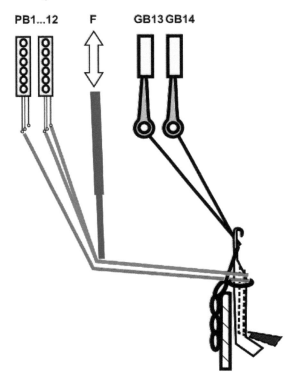

FIGURE 8.1 Guide bar arrangement with fall plate - the yarns of the bars after the fall plate are slid down to the position of the previous loops, so that they do not build loops.

Normally the ground bar moves counter lapping during the overlaps to the tuck-building bars. In this case the yarns of the guide bar cross the yarn of the tuck and it can slide down along the needle without taking with it the ground yarn. With adjustment of the yarn tension it is possible to have building of tucks with equal lapping, too. The exact form of the connection between the tuck in the loop depends on several parameters - 1) if ground and tuck guides

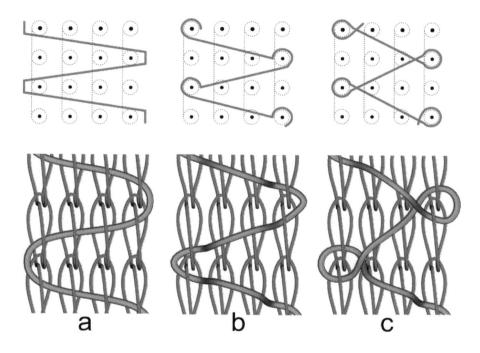

FIGURE 8.2 a) Laid in yarn in comparison to a tuck placed as a open loop b) and tuck, placed as a closed loop c). Remark: This figure demonstrates the tuck forms principally, but **not** their appearance depending on the type of the background loop and the lapping type (equal or counter-lapping)!

are moving equal- or counter-lapping during the overlapping, 2) during the underlapping, 3) if the tuck is placed as open or closed loop. In [26] the most **common** types of the tucks are described and presented for a background of tricot and lock stitch. A trial for more systematic representations was created in Figure 8.3, without guarantee, as the author was not able to test these situations practically! This figure does not cover all variants, which are listed in form of 3D matrix in Figure 8.4. From figure 8.3 it can be taken that if the tuck is placed as an open loop (Figure 8.3a, b and c) the yarn is placed in similar way as a laid-in yarn. How exactly it will build the connection depends on the direction of the underlaps of the background structure. If these are in opposite directions to the tuck yarns (Figure 8.3e) or a lock stitch is used then the tuck is kept fixed from these yarns. If the underlaps of the background loop have the same orientation as these of the tuck placing yarn (which would be the case of equal lapping Figure 8.3b but as well in the case of counter-lapping with closed loop Figure 8.3c), the tuck cannot be fixed and it start sliding into some stable position (c)Figure 8.3g), which depends on the tension and friction coefficients of the both yarns. Normally, the type of the connection makes a difference in the optics and if the same type of turning point has to be used, an open lock stitch or any other symmetric lapping (once to left, once to right) has to be applied.

Figure 8.5 shows tucks on a sample with hexagonal background. These are placed from patterning bars as elements of decoration of nice lace (Figure 8.6). Another sample with magnification is demonstrated in Figure 8.7. From both figures it can be seen that the yarns are connected within the loops only on the edges of the figures and floats between them. During the placement in some of the motion the yarn become as well small twist.

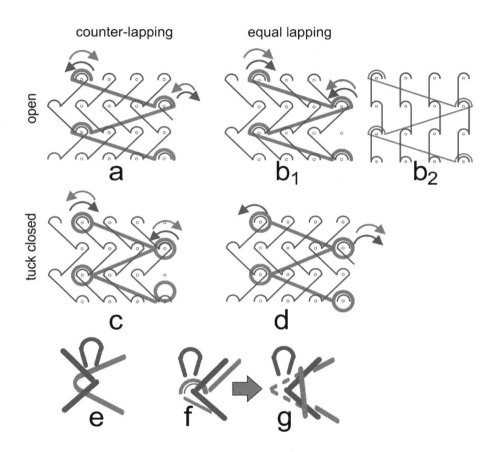

FIGURE 8.3 Combinations between ground fabrics and tucks in columns - counter-lapping (a and c) and equal lapping b) and d); in rows - type of placement of the tuck a) and b) as an open loop. e) orientation of the connection is fixed in the opposite direction oriented overlaps of the ground, f) initial state in case of equal overlapping and equal underlapping, g)position of the tuck after sliding in the case of equal under- and overlapping

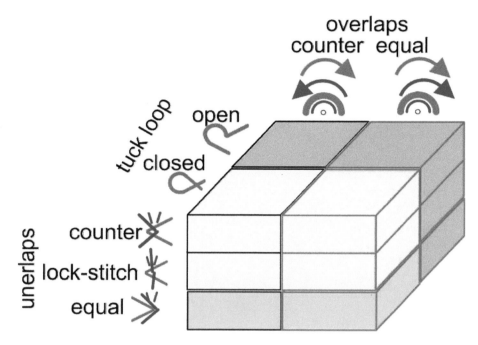

FIGURE 8.4 The combinations of the motion of the underlap, overlap and loop type gives at all 2x2x3=12 variants for tuck connections, if the lock stitch in the ground is considered as counter-lapping, these can become reduced to 2x2x2= 8 variants.

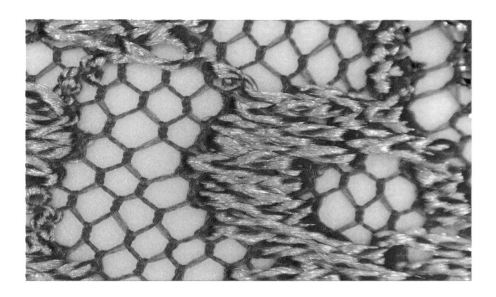

FIGURE 8.5 Microscopic image of the sample of Figure 8.6.

FIGURE 8.6 Sample E52 with tucks and laid-in elements, of the Karl Meyer company, Kettenwirkpraxis, Year 2015.

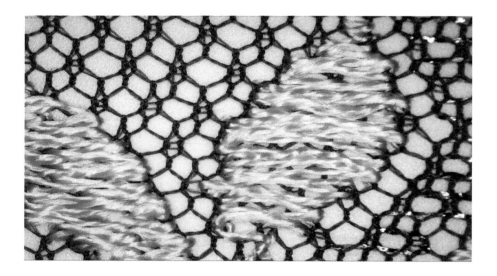

FIGURE 8.7 Another Sample with tucks, Sample 95 from the Karl Mayer company, Year 2015.

8.3 Cut presser

The (older) machines with bearded needles were normally provided with one complete plain presser (Figure 8.8a) , that closed *all* beards in the time, where the new yarn has to be pushed trough the loop head of the previous loops. With these machines it was possible to be provided with a second presser plate with special form (Figure 8.8b and Figure 8.9a), where parts of it are removed. In the cycles, in which *this* plate presses the needles, the needles against the shorter pieces remain open and the previous loops stay there, so they are miss-knitted. If these loops become knitted during some of the later cycles, they deform the structure, building three dimensional effects. These loops keep their nearly initial length (Figure 8.9d), and do not become longer as visualised principally in the figure (Figure 8.9c). The "normal" loops build several rows, which become connected through the shorter loops and build relief effects. The form of the cut presser (Figure 8.9a) and threading of the guide bar (Figure 8.9b) have to be synchronized together. Two types of pattern are possible on such machines:

- shell stitch fabrics. These are produced by only one guide bar with partial threading and suitable cut presser. The solid portions of the cut presser corresponds to the threaded guides. For these samples no plain presser is required.

- spot or knop effects. These effects require the use of two or three guide bars. The fully threaded bars work with the plain presser and the partial threaded - with the cut presser.

More details about this technique can be found in Chapter 10 of [18] or in [23] where real samples can be seen like the photographed one for figure 8.10.

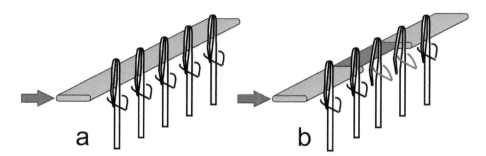

FIGURE 8.8 The cut presser.

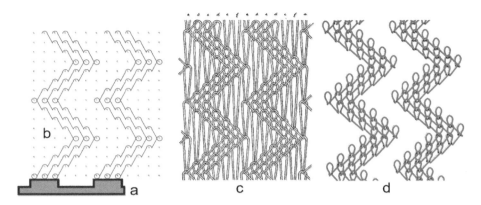

FIGURE 8.9 Cut presser fabrics a) cut presser form 3:3 b) lapping movement
c) construction, reproduced from [18], page 184 d) initial loop geometry.

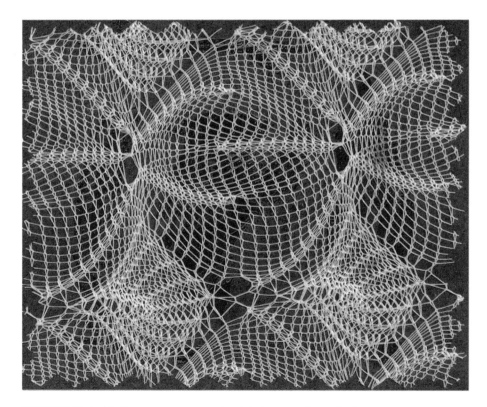

FIGURE 8.10 Cut presser sample, fabrics 137 from [23].

8.4   Pleating

The individual, controlled feeding of the yarns of some of the warp beams allow creation of pleats and other 3D structural effects. The basic structure is built from one, two or more guide bars (Figure 8.11b,c) and is a normal plain structure (Figure 8.11d). One additional guide (Figure 8.11a) builds loops and then do not overlap needles for several cycles. If the yarn feeding is not adjusted, the yarn will be straight (Figure 8.11f), but if the yarn feeding for this beam only during the misslapping is reduced, then the remaining structure will build pleats, arranging itself in the thickness direction. Microscopic view of such sample, shown in Figure 8.12, demonstrates the fine pleats.

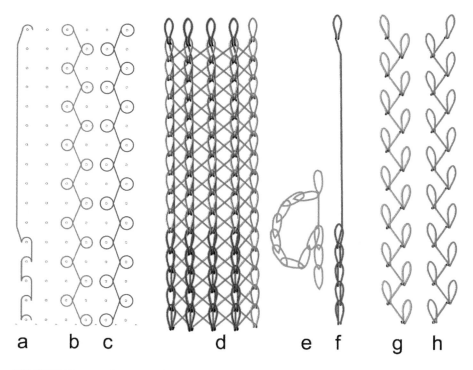

a   b   c   d   e   f   g   h

FIGURE 8.11 Pattern for pleated stucture. a) misslapping guide bar b) and c) lapping for the ground structure d) simulation of the ground structure, e) relaxed form of the loops if the misslapping yarn has a shorter length f) g) and h)- principle 3D simulation of the lapping for each bar without the length adjustment.

FIGURE 8.12 Microscopic view of pleated sample from [23], p.111.

8.5 Electronic beam control

Modern machines can have individual, programmable drives for the take-off beam (EAC = electronic Abzug (German word - take-off) control) and for the warp beams (EBC = electronic beam control). The computer controlled drive on a take-off beam allows production with different take-off speeds in different regions of the pattern and is known as "multi speed". All loops in one course will have shorter height if the speed (for the same time step) is smaller. Programming the different loop heights per cycle allows designing of a pattern with regions of different density and appearance (Figure 8.13) only based on the loop size.

The individual control of the yarn beams allows perfect control of the yarn *length and tension*. For the production of samples with miss-stiches for pleated fabrics the yarns have to be separated into sections and each section is placed on separate beam. In this way the length and the tension of the yarns in each section can be controlled individually. Additionally to the pleated effects (Section 8.4), different structural relief patterns can be produced. Figure 8.14 gives only one idea for combinations of areas with different yarn tensions. The higher tension in some regions would lead to relaxation of their loops, making them shorter. The longer neighbour regions change, then their plain form drapes into some 3D relief.

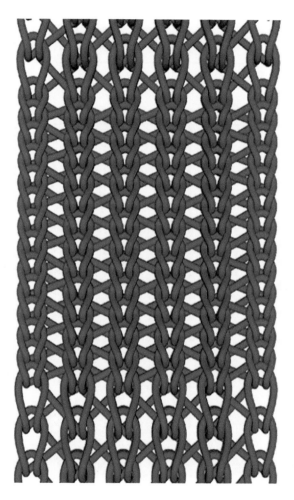

FIGURE 8.13 Modern computer controlled take-off drives allow production of regions with different loop heights.

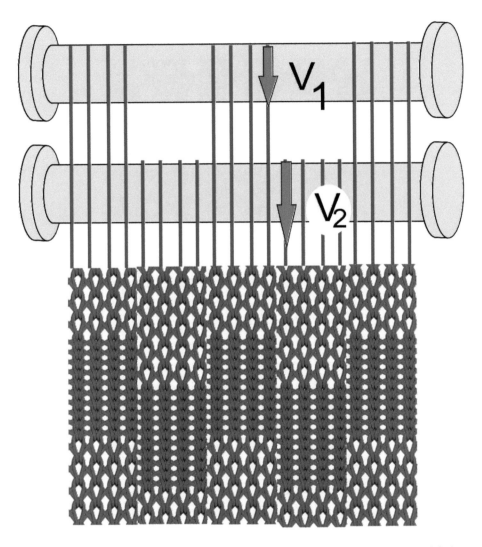

FIGURE 8.14 Principal patterning of sample with section beams with individual control. Different regions can be produced with different tension and can relax after that to shorter loops. The structure will have 3D relief after the relaxation.

8.6 Conclusions

The fall plate technique extends the patterning possibilities with nice stitching like effects, where the yarn floats and remains connected only at the edges. The placement of the yarn at the turning points depends on the combination between the lapping directions of both tucks and loops. The cut presser technique allows nice samples with shell or spot effects, but these techniques can be applied mainly on older generation machines with bearded needles. The electronic beam control extends the patterning with a possibility for having loops with different sizes and for patterns with 3D surfaces or pleats in the case of sectional threading.

9

Jacquard fabrics

9.1 Introduction

The invention of the Jacquard machine in weaving allowed effective individual control of each warp yarn in weaving. The idea was transferred to warp knitting, where single guides are individually controlled in one guide bar (Figure 9.1). In case of mechanical or electric command, each guide on the jacquard guide bar can be bent so that it places the yarn at one position higher then the original one in relaxed, or not activated position. Currently the most advanced are the piezo-jacquard bars [15],[16], where at each guide one piezo-electric transducer is fixed. The transducer deflects the guide by application of a control potential. Information about the older mechanical or electromechanical systems can be found in a several books. This chapter concentrates on the fabric construction of these fabrics. The possibility to move each guide individually opens enormous patterning possibilities in warp knitting, because it increases the variations of the loops without the need of a large number of single bars. From another point of view, the jacquard bar is more costly and requires special software and good educated people for the design of the fabrics. For these reasons warp knitted machines with jacquard guide bars are used mainly for products where individual local changes of the pattern are required.

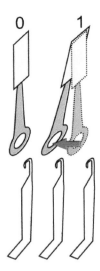

FIGURE 9.1 Each jacquard guide is individually controlled and in case of command (1) can be bend so, that lap at the next one (+1) position.

9.2   Jacquard types and machine configurations

With the jacquard guide bars patterns based on three elements can be produced:

- loop underlaps with different lengths, because of the changed loop position;

- laid-in with different lengths;

- tucks underlaps with different lengths.

The activation of one guide at the jacquard guide deflects this guide at one position and makes its loop underlap, laid-in or tuck underlaps - longer or shorter- depending on the basic motion of the guide bar and the time of the activation. If the element is longer, it crosses more then one wale and produces **a denser area** in the fabrics. If the element is shorter - the area is **more open** or not even connecting eventually available ground elements.

For production of underlap based jacquard the jacquard guide bar(s) are placed as first bars (Figure 9.2a) and build normally loops. If the jacquard guide bar is placed as the last guide bar (Figure 9.2b), it has the best position for placing laid-in yarns fixed between the loops. A jacquard guide bar at the first position but before the fall plate (Figure 9.2c) produces tucks.

There is another common configuration, which can be considered as an extension of the configuration (Figure 9.2a). For this case the jacquard guide bar is responsible for the building of loop based jacquard ground fabric, on which tucks are placed additionally for decoration (Figure 9.3) using pattern bars after the fall plate. Each of these configurations has its advantages and disadvantages and specialized application areas. The placement of laid-in yarns is probably the simplest for design and most stable for the knitting process of these three. It allows filling of different areas, using thicker yarns. The laid-in yarns are connected between the loop legs and underlaps of the ground bars.

The loop based jacquard can be applied only for yarns with usual for the machine fineness, not for thicker yarns, because they build loops. It is commonly used with hyperelastic yarns or for fabrics where the different densities are sufficient to represent the desired design. Tuck based jacquard requires a machine with a fall plate. Because the tucks do not build loops, here again thicker and cheaper yarns can be used for effects.

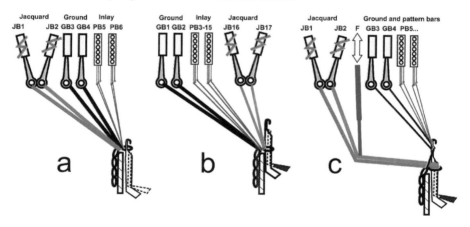

FIGURE 9.2 Three types of jacquard warp knitting machines.

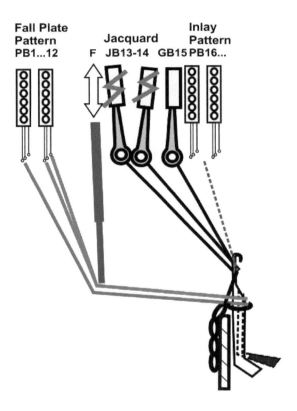

FIGURE 9.3 Jacquard bars for the ground and fall-plate design too.

9.3 Principle of loop based jacquard

Loop based jacquard fabrics are using machine configurations like those presented in Figure 9.2a. The jacquard bars are placing their loops not always on the the same needle, and the exact pattern depends on the design. In order to ensure the continuous knitting process of each needle, additional ground guide bars are placed. These provide some basic fabrics and guarantee that once at one needle there is a loop, there will be such in all following cycles. The jacquard bar has a lapping motion normally based on tricot stitch (Figure 9.4). If no one guide is activated, all guides will build the tricot stitch. Each guide can be activated and deflect during the swing-in step and during the swing-out step. In the example of (Figure 9.4) the guide is deflected only during the first swing-in steps within a repeat of two courses. For this single guide the lapping motion changes in this way from 1-0/1-2/ to 2-0/1-2/. Mathematically, for the two knitting cycles, which build a tricot stitch, there are sixteen possible configurations, which can be created, and which are demonstrated in Figure 9.5 and Figure 9.6. Some of them lead to placing yarn without connection, as the configuration Figure 9.5 g, because this is laid-in lapping, but the guide bar does not have a suitable position for placing laid-in yarns. Normally, for one machine and one basic pattern only a few configurations are selected, which can represent visible different densities.

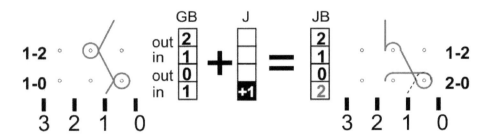

FIGURE 9.4 The lapping movement of each guide (not guide bar) is the result of the motion of the guide bar and the individual signal of piezo- or mechanical system for the guide deflection motion.

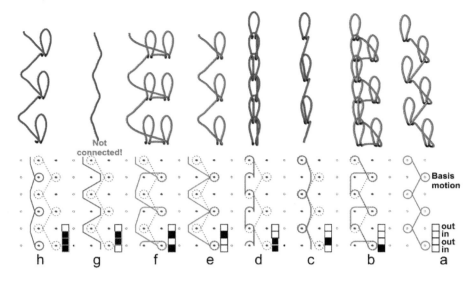

FIGURE 9.5 First eight theoretical laps based on tricot lapping.

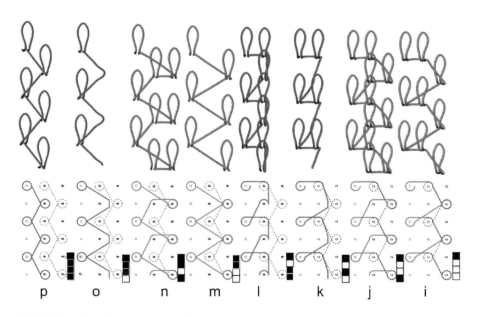

FIGURE 9.6 Second eight theoretical laps based on tricot lapping.

9.4   Design colours

During the design of jacquard products, the designs create a painting with different colours. Each colour in this painting corresponds to an **area** with some structural effect or density or colour appearance. In the case of warp knitted structures the smallest area could be a single loop, if plating effects are applied, but in a normal case a sequence of two knitting cycles (one repeat) for one needle is used. During the two knitting cycles the needle places an underlap in the one and then in the opposite direction and creates one repeat. The design colours are translated in a second step into a knitting pattern. Let a ground tricot structure be created by one ground guide bar and let the jacquard bar have tricot basic lapping, too. The underlaps of the ground tricot structure are visible in Figure 9.7 in the 3D simulation with "Thinnest effect". The thicker loops are the plated loops based on the lock stitch of the jacquard bar, with pattern Figure 9.7d. These are created by activation of the guide during both swinging motions in the **first** knitting step. Such structure is transparent and is marked with white colour in the design [5],[14]. The normal, basic motion of the guide in the jacquard bar will remain a tricot stitch and will place underlaps as well. If the jacquard yarn is thicker, then the ground structure, a medium transparent structure will be created. This is marked with green colour. If the second swing-in and swing-out steps in the second cycle are activated, the tricot stitch of the guide will be modified to cord stitch (Figure 9.7m). For this case, **two** underlaps will cross the empty space between the loops and will fill better in this space. The less transparent structure is as well thicker, and the areas with it are marked with red colour. The remaining cases (Figure 9.7h and c) can create similar effects like the (Figure 9.7d), but will not build loops with each second cycle. If the guide is deflected all the time during one cycle, (Figure 9.7p) it will create tricot again, but with the next two needles, and the space, related originally to this needle, will remain empty. In a normal case the cases d) a) and m) of Figure 9.7 are applied, the other can be assigned to different colours for special effects.

Figure 9.8 demonstrates a simple rectangular mesh, where the different areas are marked with the different colours. The size of the mesh cells depends on the fineness of the machines and the take-off speed during production. Once this design is available, the CAD software today is able to transfer it automatically to a pattern. Figure 9.9a demonstrates the design created with the TexMind Pattern Editor lapping diagram for the jacquard guide bar only and on the right hand side - Figure 9.9b demonstrates the 3D simulation of the loops of the jacquard guide bar. The careful reader will detect some missing loops on some needle positions. These **has to** be filled by the yarns of the **ground** guide bar and are not presented in here in order to keep the figure clear. If the basic tricot lapping of the guide bar starts with one cycle later (instead of 1-0/1-2// it becomes 1-2/1-0//) the resulting effects of the

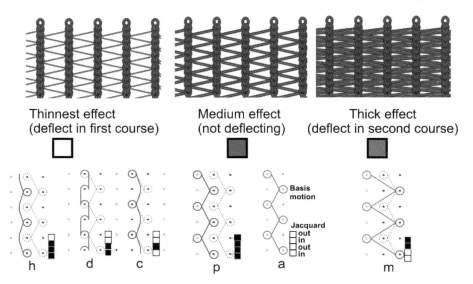

FIGURE 9.7 Left lapping (h, d,c) builds thin, or transparent structures, the basis tricot (p) and (a) builds medium density and the right structure (m) builds longer underlaps and results in a solid structure. Common colours for the patterning of these three structures are white, green and red.

white, green and red colours will be changed and become almost mirrored, as visualised in Figure 9.10. The denser red area becomes almost transparent, the green area remained green with normal tricot and the transparent white area becomes denser (red). If for the basic motion of the guide bar is taken cord stitch, instead of tricot stitch, there will be already three underlaps in one area and all three colours will produce a denser appearance (Figure 9.11). The cord stitch as a basis allows some more effects, for instance if overlapping over two needles is used, and using it some more design colours can be applied. Some more patterning possibilities are available if two jacquard guide bars are used with counter-lapping (Figure 9.33), as presented in the last section of this chapter.

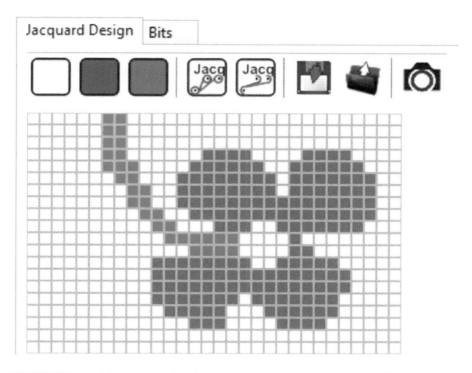

FIGURE 9.8 The jacquard colour pattern design marks the different areas, where the different densities will appear. The colour areas are transferred to single bits in the modern CAD software automatically. The screenshot is of the TexMind Warp Knitting Pattern Editor.

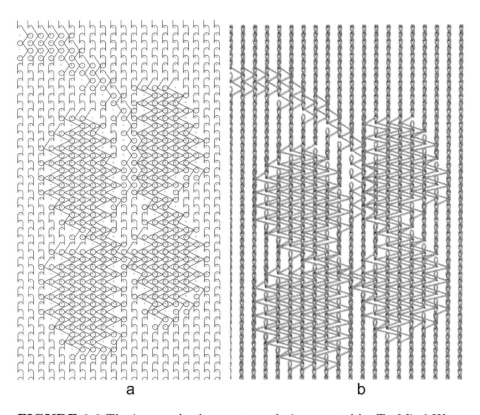

FIGURE 9.9 The jacquard colour pattern design created by TexMind Warp Knitting Editor for the design of Figure 9.8 a) lapping diagram for the jacquard guide bar b) simulation of the loops of this guide bar.

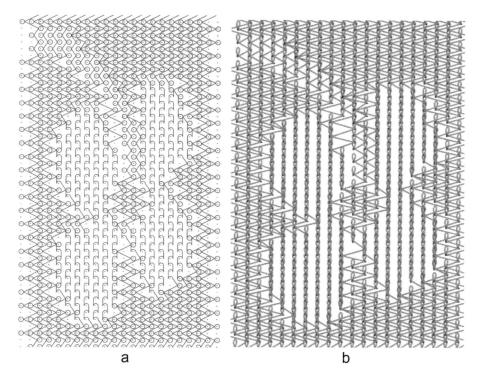

a   b

FIGURE 9.10 The pattern of 9.9, in case the basic tricot motion step is moved one cycle up. a) lapping diagram of the jacquard bar, b) 3D simulation of the loops of this guide bar.

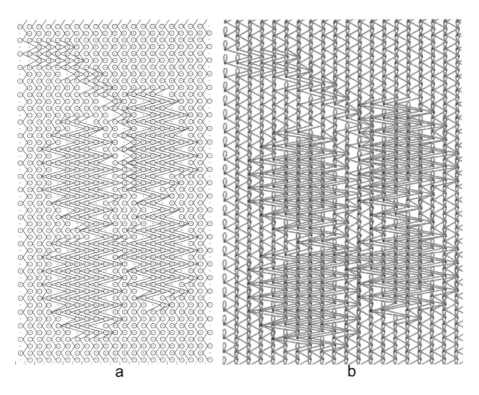

a   b

FIGURE 9.11 The pattern of Figure 9.8, transferred for guide bar with basic **cord** stitch, instead of tricot.. a)lapping diagram of the jacquard bar, b) 3D simulation of the loops of this guide bar.

9.5 Fabrics sample

An example of jacquard fabrics is demonstrated in Figure 9.12 as larger overview and in Figure 9.13 with zoomed in some detail areas. The open background A) becomes a little be more gray or dense, if the some underlapping appears at area C) or very dense in the well connected areas B). The underlaps are well recognized on the back side of the fabrics, Figure 9.14 and Figure 9.15. This sample is produced on one RSJ4/1 machine of the the Karl Mayer company, Holding GmbH & Co. KG, Obertshausen, Germany, with gauge E28 and five guide bars. The chain notation for the guide bars are: Jacquard bars JB1: 1-0/1-2//, JB2: 1-2/1-0//, GB3: 1-0/1-2/2-1/2-3/2-1/1-2//; GB4: 2-3/2-1/1-2/1-0/1-2/2-1//, GB5: 0-0/1-1//, where GB3 and 4 are with half threading and guide GB5 with full threading, building power net as a ground fabric. This is only one example of such structure, the combinations of materials, ground fabrics and machine settings allow a very large number of effects.

FIGURE 9.12 Part of sample 2012/0048 of the Karl Mayer company, produced on RSJ 4/1 machine.

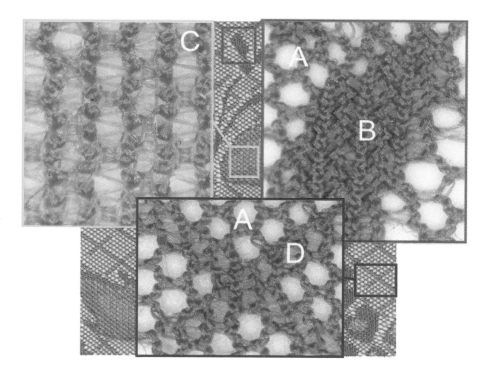

FIGURE 9.13 Three characteristic areas in the jacquard sample -a) open background, b) dense structure (c) and (d) built paths with middle density.

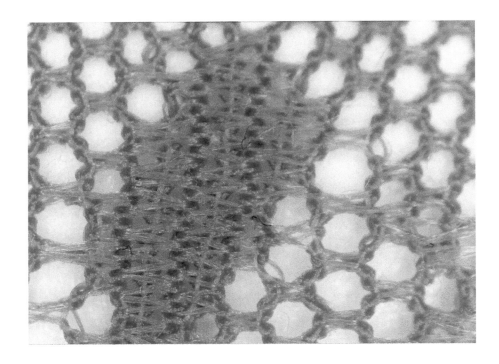

FIGURE 9.14 Underlap side of the solid region B) of Figure 9.13.

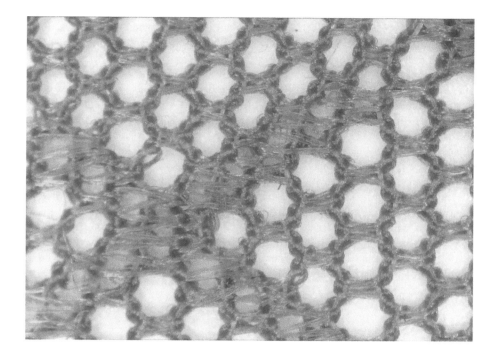

FIGURE 9.15 Underlap side of the solid region D) of Figure 9.13.

9.6   Laid-in based jacquard

For the laid-in based jacquard the jacquard guide bars have to be placed last on the machine, or at least behind some ground bars, which will build holding underlaps (Figure 9.16). The ground bars GB1 and eventually GB2 have to build the ground structure. Some more colour effects can be placed by patterning guide bars. The jacquard bars in this case place laid-in yarns between the loops and change again the connectivity between the single wales and the filling degree of the area between the loops. As basic motion here is selected, similar to tricot stitch in the loop based jacquard, a underlapping motion under two needles 0-0/2-2//. The white area in this case shows only the ground structure, which in Figure 9.17a is a lock stitch. All guides in this area are deflected during the first step, so that the equivalent lapping is then 0-0/1-1// and this is laid-in around one chain, without connection to the next one. For the green area, Figure 9.17b, no deflection of the guide is planned, it performs the normal motion and places laid-in yarn between two wales. Deflecting during the second cycle leads to 0-0/3-3// lapping, which leads to longer laid-in, crossing two chains. The two visible laid-in yarns between the loops, Figure 9.17c, fill the space better and make a darker and denser appearance in this red area. The rules for a number of the interlacing points from Chapter 7 remains valid here and have to be considered. For instance, the lapping motion of the ground bar and the pattern bar (jacquard) during the overlapping has to be in the opposite direction. With this counter-lapping the laid-in yarn remains fixed at the placed position and does not slip to the next position. The pattern of Figure 9.8, transferred using the here described principle, is presented as a lapping diagram in Figure 9.18a, simulated yarn position of the jacquard bar only Figure 9.18b, and with the simple ground of lock stiches Figure 9.18c.

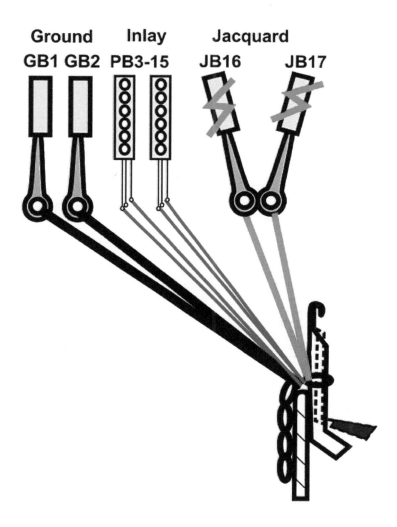

FIGURE 9.16 Guide bar arrangement for laid-in based jacquard.

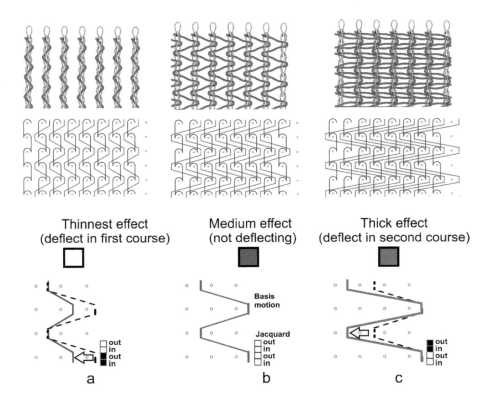

FIGURE 9.17 Laid-in based jacquard - principle regions a) white area with visible only ground structure, b) green area with one laid-in visible c) red area with two laid-in yarns between two loops.

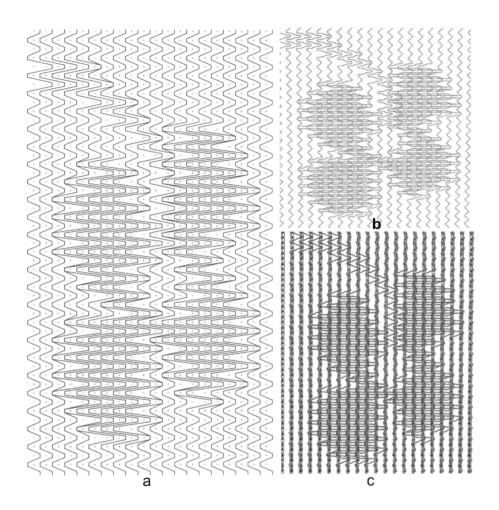

FIGURE 9.18 Laid-in based jacquard structure for the design of Figure 9.8. a)lapping diagram for the jacquard bar, b) 3D simulation of the yarn placement of the jacquard bar c) 3D simulation of the jacquard bar and lock stitch based ground structure.

9.6.1 Laid-in jacquard examples

Figure 9.19 shows one possible ground structure, produced of one ground bar and two jacquard bars. The jacquard bar is placing thicker laid-in yarns between two needles and forms the basic mesh. If at single steps one guide gets deflected to a +1 position (Figure 9.20 path J), the area become filled by the laid-in yarn. In this way areas with different pattern, including characters can be produced, as the view of the piece of this fabrics demonstrates (Figure 9.21)

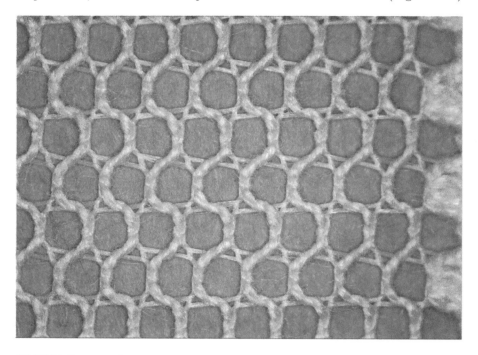

FIGURE 9.19 Ground structure of laid-in based jacquard. Sample J/3391 of the Karl Mayer company, designed by Schörner.

A similar technique with synthetic yarns gives other optics of the sample. Figure 9.22 presents a transparent area based on one guide bar loops and laid-in in the jacquard bar, where the laid-in makes one additional deflection at each second cell. In this way, only the *half* of the cell is filled. Two deflections within one mesh cell (Figure 9.23) makes it denser, and with participation of the yarns of the both full jacquard bars on this machine the denser areas can be formed. In the view of the complete sample (Figure 9.24) mainly two areas become recognized optically, but from larger distance the area with the pattern of the Figure 9.22 starts to appear too.

Using laid-in allows the use of thicker and voluminous yarns which increases the possible combinations of the patterns and gives the art-designers more freedom in the patterning. Using for instance staple fibre yarns - in this

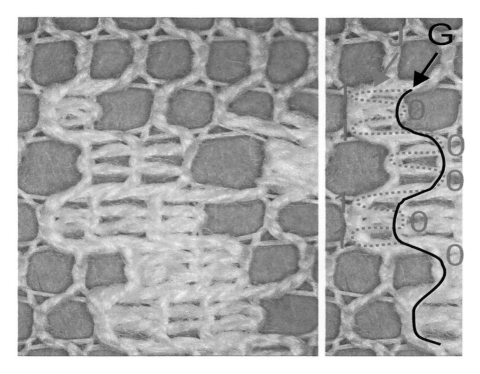

FIGURE 9.20 Jacquard effect is produced by deflecting the guide. G- yarn path without deflection, J - resulting yarn path based on the deflection at single steps. Sample J3391 of the Karl Mayer company, designed 1982 by Schörner.

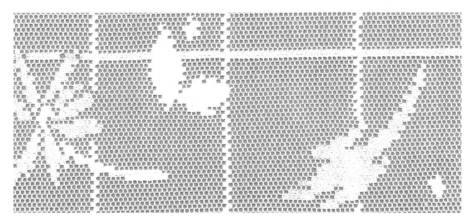

FIGURE 9.21 Complete structure. Sample J/3391 of the Karl Mayer company, designed by Schörner.

FIGURE 9.22 Laid-in based jacquard pattern for filling of transparent area.

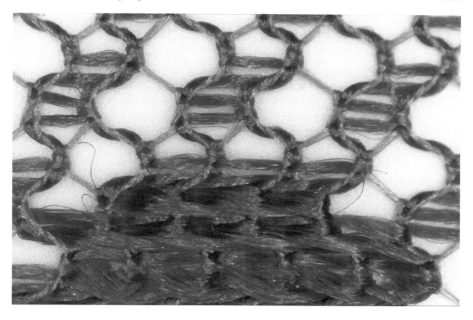

FIGURE 9.23 Laid-in based jacquard pattern for filling of denser areas and the dense design part.

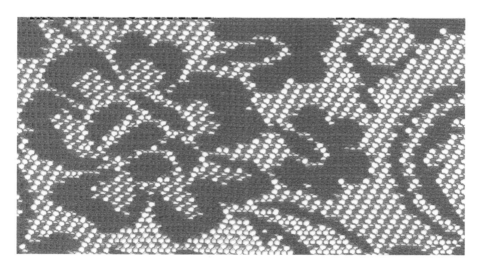

FIGURE 9.24 Sample J/3141 of the Karl Mayer company with regions of Figure 9.22 and Figure 9.23.

case 50% polyester/50% cotton 20 tex x 2 (Nm 50/2) with ground of 100 dtex f40 polyester makes a sample untypical for warp knitted fabrics' stable fibre look and touch. The production of fabrics with staple fibre yarns requires cleaning of remaining fibres and dust, but is generally possible on suitable machines, in this case with gauge E14. Figure 9.25 demonstrates filled area of such sample, Figure 9.26 the ground mesh and Figure 9.27) the complete sample, produced in year 1983. Both of these samples demonstrate the large patterning possibilities of the jacquard technique with a minimal number of guides.

FIGURE 9.25 Laid-in based jacquard with lock stitch ground and laid-in weft. Filled area.

FIGURE 9.26 Laid-in based jacquard with lock-stitch ground and laid-in weft. Ground fabrics.

FIGURE 9.27 Laid-in based jacquard with lock-stitch ground and laid-in weft. Complete sample J/3687 of the Karl Mayer company, developed by Schörner.

9.6.2   Laid-in jacquard with different ground pattern

The jacquard guide bars which place laid-in yarn can be used often to create different backgrounds in the fabrics. The combination of this technique with laid in of normal (string) guide bars, as demonstrated by the laces, allows large variation of the optics in the different areas of the same sample. Figure 9.28 shows four different ground types, within the same sample (Figure 9.31). The densiest filling is if the laid-in yarn connects two wales each second cycle (Figure 9.28D). If this connection is not done every second type, but every fourth cycle, the mesh becomes more transparent (Figure 9.28D). Larger regions without connection between the wales causes side inclination of the wales in the contact areas, which in the case (Figure 9.28B) leads to hexagonal type of even more transparent mesh. Hexagonal mesh with thicker boundaries can be built if **two** wales become continuously connected into one chain and connects only on after certain cycles to the next such chain (Figure 9.28A). The connection of two chains together into one builds larger openings around (Figure 9.29D) and in case both laid-in yarns works together thicker single chain (Figure 9.29A). The design areas can be filled from the patterning guide bars (not jacquard, string bars with single guides) with the same type of pattern but with yarns with different linear density and creates in this way an additional possibility for getting different appearance (Figure 9.29B and Figure 9.29C and Figure 9.28F). In order to get better optics of the figure boundaries, the yarns on the edges are placed on separate guide bars and placed as thicker laid-in on the edge only (Figure 9.28G. The edge yarns can be used additionally to create separated figure effects Figure 9.29E). During the production of underwear and decoration tapes the samples are cut to some nice, not orthogonal profile. The connecting area for the current sample is marked in Figure 9.30. The complete sample, demonstrated in Figure 9.31 was designed and produced in early 1982 on a machine with a total of 56 guide bars. The ground yarns are dtex 33f1 polyamide, the jacquard guide has dtex 44f10 polyamide, the pattern bars are with thicker dtex 156f34x2 and dtex 156f34x1 textured polyamide. Another beautiful laid-in jacquard sample for curtains is presented in Figure 9.32.

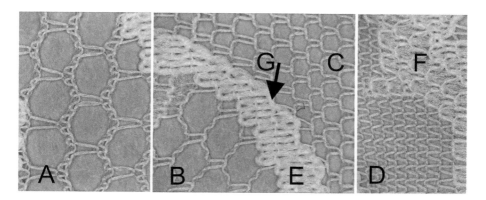

FIGURE 9.28 Different laid-in based jacquard ground pattern. A) based on two wales, B) one wale C) connecting two wales every fourth cycle D) connecting two wales each second cycle F) and E) are design areas, G) boundary yarn.

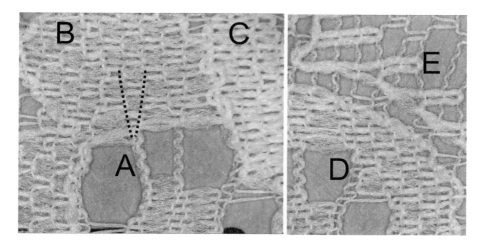

FIGURE 9.29 Laid-in based jacquard based on lock-stitch ground. A) two wales connected together for making a larger opening D) not so hard connection between the two wales, B) filled design area with dtex 156f34x1 C) filled design areas with the same pattern as case B but double thicker dtex 156f34x2 yarn for better filling; E) edge decoration.

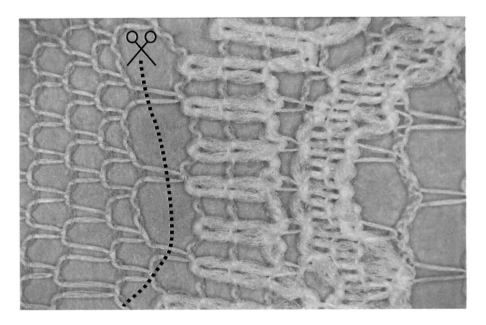

FIGURE 9.30 Laid-in based jacquard based on lock-stitch ground. Sample with no rectangular edges. The cutting line is marked.

FIGURE 9.31 Laid-in based jacquard based on lock-stitch ground with 56 Guide bars. Sample SJ56/001 of the Karl Mayer company, designed by Oppolzer.

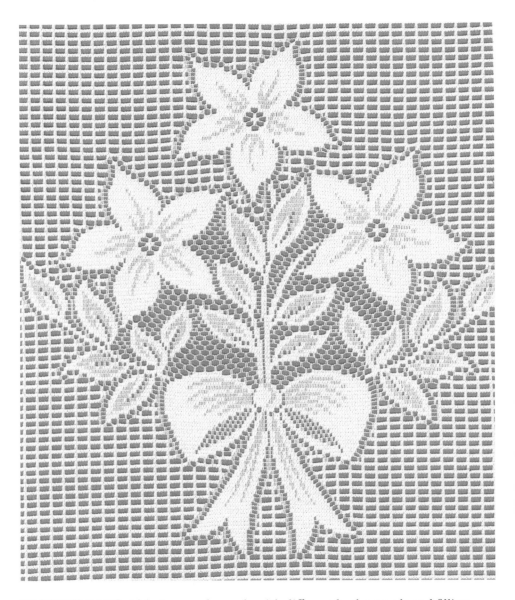

FIGURE 9.32 Laid-in jacquard sample with different backgrounds and filling areas.

9.7 Tuck (fall plate) based jacquard

Tuck based jacquard follows the same principle as the previously described structures. The warp knitting machine has to be equipped with at least one guide bar for production of a ground structure and can have additional pattern or guide bars (Figure 9.33). The jacquard bar is then placed as in first position and the fall plate moves down its yarns, after the overlapping, so that they do not build loops. In [5] are reported four types of regions, build analogously to the four loop-based or laid-in based regions - with larger and shorter floats. This technique is reported in several texts in the literature and there are known older machine configurations with such arrangements. Today this technique is used mainly if stitching similar effects have to be produced, where larger (not connected within the loops) floats with individual placement are required.

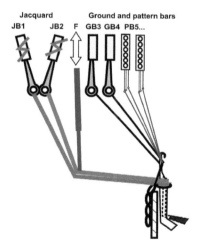

FIGURE 9.33 Machine configuration for tuck based jacquard. The machine required fall plate and jacquard bars which are before it.

9.8 Tuck (fall plate) with jacquard ground

The fall plate effect can be applied with samples of different jacquard ground. The jacquard guide bar is used in this case in a similar way as in sample in Figure 9.31 to arrange areas with different densities and optics and the bars after the fall plate places additional, more visible yarns with larger floating

yarns. Figure 9.34 presents detailed views of different regions of the sample of
Figure 9.35, which is produced in this way. The two main background regions
Figure 9.34 b and Figure 9.34 c are produced based on lock stitch and laid-
in yarns on the jacquard with different connections between the wales with
similar pattern to those of Figure 9.28. The tucks (Figure 9.34 a) build longer
floating areas, placed from pattern bars and are the main effect of this sample.

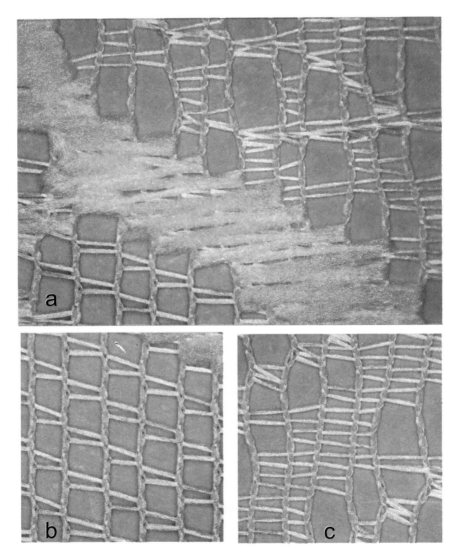

FIGURE 9.34 Two ground areas with laid-in jacquard and patterning area
with tucks from pattern bar after the fall plate.

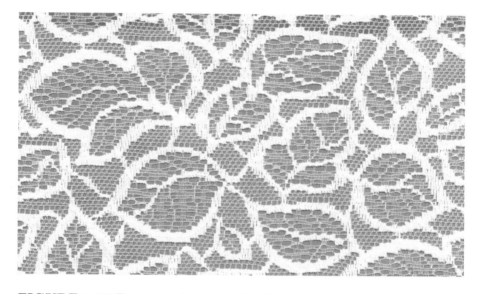

FIGURE 9.35 Two ground areas with laid-in jacquard and patterning area with tucks from pattern bar after the fall plate, complete sample Nr 28007/727 of the Karl Mayer company, designed by Mahr.

9.9   Threading and control of jacquard bars

Each guide in one jacquard guide bar can deflect and tape two positions - its normal position and the next one. For this reason, the next position has to be empty and one jacquard guide bar has half the number of guides from those of one normally fully threaded guide bar. For this reason, normally two jacquard guide bars are working synchronously as one shog line (Figure 9.36a). The second jacquard guide has its guides placed at one position later but both of these have the same ground lapping motion. Such one combined jacquard guide bar is noted in the documentation as JB1+2, and during the development of its pattern can be considered a one fully threaded jacquard bar. The distribution of the commands to the single guides is normally performed automatically from the CAD software. For some samples, the second jacquard guide bar can move counter-lapping to the first one (Figure 9.36b). With machines of such configuration the sample is prepared, reported in Section 9.5. Figure 9.37 demonstrates some possibilities for the common work of the two counter-lapping jacquard guide bars. If the first loop of the red guide and the second loop of the blue guide are moved to +1 position (Figure 9.37a), then a plated lock-stich chain will be built at one of the needles and the other needle will have no yarns from the jacquard guides. This corresponds to a very transparent area. In the normal state, without deflection of the guides, they will produce tricot chains (Figure 9.37b), which correspond to a normal transparency or normal filling degree, marked as green colour. If the second loop of the red guide is moved to +1 position (Figure 9.37c) only, then its yarn will have a longer underlap and will make a denser filling. Moving the first blue loop and the second red loop to +1 position will turn the tricot stitch for both guides to cord stitch and will lead to longer crossing underlaps with even denser filling (Figure 9.37d).

Five regions with colours: white (Figure 9.38a) for transparent, yellow (Figure 9.38b) for semi-transaperent, blue (or black) for dense filling, green for transparent thicker vertical chains (Figure 9.38d) and red for very dense structure (Figure 9.38e) using overlaping over two needles are reported in [9].

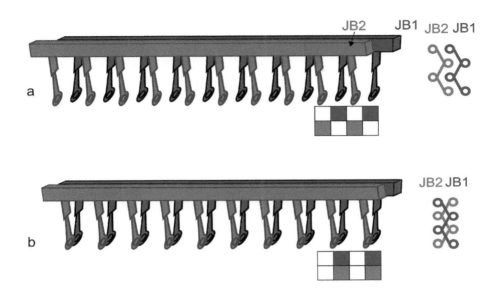

FIGURE 9.36 Possible configurations of two jacquard guide bars a) with equal lapping as one equivalent fully threaded guide bar, b) as two half threaded counter-lapping bars.

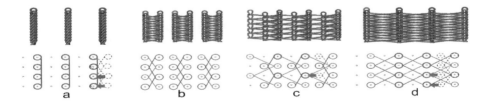

FIGURE 9.37 Possible configurations for counter-lapped jacquard bars a) both tricot stitches transferred to lock stitch, b) normal configuration, c) one of the tricot stitches extended to cord stitch for longer underlaps d) both tricot stitches extended to cord stitch for longer crossing underlaps.

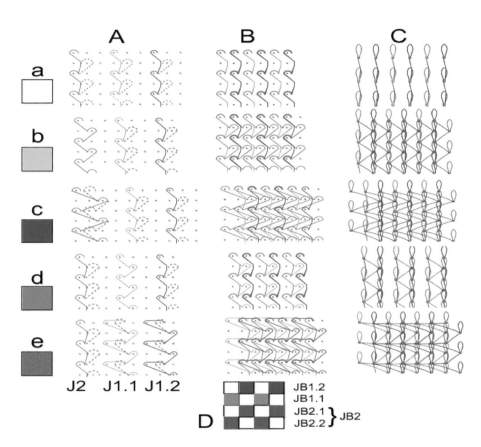

FIGURE 9.38 Jacquard patterning with two full jacquard bars. The two jacquard guides J1.1 and J1.2, which build one fully threaded set can change their pattern at different times and in combination with the full set of JB2, five total structural elements can be obtained.

9.10 Conclusions

The jacquard guide bars can be placed at different positions on the machine and depending on that position, can be used to create local effects with different structural elements - underlaps of loops, underlaps of tucks and laid-in yarns. The variety of patterns in such cases is very large and seldom used. The main application areas are based on tricot or cord stitch motion of the guide bar (respectively 0-0/2-2// for the laid-in variant), and the guides in some areas are deflected during the first or second cycle in order to shorten or extend the length of the underlap and in this way to change the filling degree of the space between the loops. Using these configurations, the designer can create in short time machine files with suitable CAD software for samples with very large repeats and interesting local effects. More complicated structures and effects are possible and require a good understanding of both of the patterning principle and the machine controls for the given machine. The best way to learn such jacquard patterning are the courses and direct communication with the machinery producers and their software vendors. Only they have the complete information about the adjustment of the machines and the settings in the software, which works with such machines. As the variations in this area are very large and the experts try to keep their knowledge from competitors, there is actually no *modern* public literature available in this area, known to the author of this book.

Part IV

Structures from double needle bar machines

10

Double needle bar structures - fundamentals and double face structures

10.1 Introduction

The availability of a second needle bed on one warp knitting machine extends significantly its patterning possibilities. Most of the patterns of all previous chapters can be produced on the **front** needle bed and combined with some of **other** patterns on the back needle bed. The number of these combinations is already enormous. Additionally to the combinations, on the machine with two needle bars, additional structural elements - connections between the beds are possible, too. This chapter presents the principles of the coding of the double needle bar samples, specific issues about the loop type and connections and the basic stitches. Finally, structures with one guide bar are presented. The samples with more guide bars are presented in Chapter 11. In most cases the samples are simulated with a larger distance between the beds in order to improve the visibility of the separated elements. For denser fabrics this visualised artificial distance has to be ignored by the reader.

10.2 Machine configuration

The double needle bar machine has, as its name says, two needle bars and several guide bars Figure 10.1. These machines are known as (double needle bar) **Raschel** machines. One complete knitting cycle consists of two subsequent knitting cycles of each of the needle bars, which makes four lapping motions. The guide bars swing-in between the needles from the front needle bed, overlap the yarns over its needles (Figure10.1 a), and then swing-out between the needles and these move down to finalize the knitting cycle. After this half double cycle, the guides swing-in between the needles of the back needle bar, overlap them (Figure10.1 b), swing-out and the back needle bed builds its loops. In some machine configurations the farthest front and farthest back guide bars work only with their needle bars and do not reach the other bed. The distance between the needle bars FA is a very important parameter of the machine. It determines (together with the lapping!) the length of the connecting yarn pieces and in this way the thickness of the fabrics. The construction and the dynamics of the normal double needle bar machine and this of high-distance (spacer) machine differ in several elements, but the principle of the loop formation remains the same. For this reason both the spacer fabrics and the double face fabrics are considered together. The guide bars in the double needle bar machine are numbered starting from one in the front guide bar.

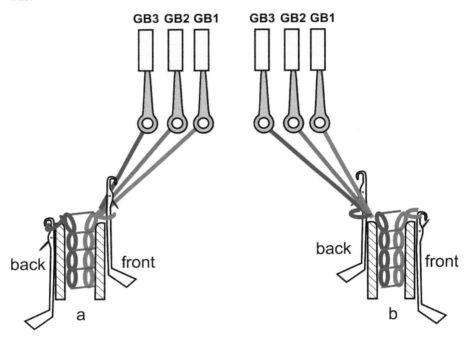

FIGURE 10.1 Double needle bar machine with three guide bars. a) Overlapping of the front needle bar b) overlapping of the back needle bar.

10.3   Notation of the lapping movement

The notation of the lapping movement of the guide bars is based on the notation of single needle bar machines, with some extensions. Each uneven cycle (1st, 3rd etc.) is assumed to present the lapping over the needles of the **front** needle bar. In order to recognize its points more easily, these are drawn larger (Figure 10.2c. and d). If such paper is not available, it is usual to be given the advice **F** for front bed and **B** for the back needle bed cycle close to the beginning courses. In order not to overfill the screen, in the TexMind Warp Knitting Pattern Editor, the character "b" is visualised only for the back course. Another issue for improvement of the readability is the use of the separating slash "/" only **after the complete double cycle**. In this way the lapping 1-0/0-0/ has to be noted as 1-0-0-0/ on the double needle bed machines.

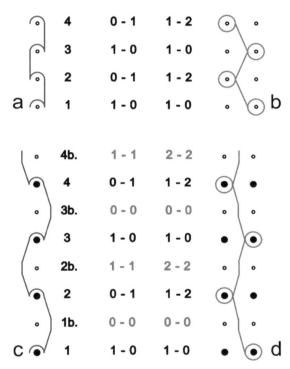

FIGURE 10.2 Lapping on single and double needle bar machines, a) and b) of single needle bar machine c) and d) the corresponding laps for production of the **same pattern** on double needle bar machine.

10.4 Design principle of double face fabrics

All laps for single needle bar machines can be transferred and applied to a double needle bar machine. Of course, the production of single face fabrics on double needle bed machines is not economically feasible, because only half of the cycles will be used. During each even (or each uneven) cycle, the guides have to miss-lap with the yarns, so that no loops are built there. The additional positions are marked red in Figure 10.2c) and d) for lock stitch and tricot stitch. This transformation process of one lapping is not only required if a single face fabric has to be created, it is the main developing step of the double face structures. If a designer or an engineer has one idea for the fabrics' appearance for the both front and back faces of the fabrics, then the laps for these two, not completely, but almost independent, face fabrics can be done separately and then merged together. The laps for the front bar are placed in the points for the front bar, the laps for the back needle bar are transferred to the "back" positions". These laps can be connected and performed with a single guide bar, or two or more independent guide bars - this depends on the type of the connection between the both faces. Figure 10.2 demonstrates this process for three separate guide bars and Figure 10.8 for a double face fabrics produced of a only one guide bar.

Guide bar A in Figure 10.3 laps in tricot lapping only over two needles of the front needle bar only. During the second half cycle the guide bar swings in and out between the needles of the back needle bar at the same position (0-0, then 2-2) and does not build any loop there. This leads to a tricot fabric for the front face. Bar B makes the same motion but with a shift to one cycle up, so that the loops are placed only on the back needle bar. Only Bar C makes loops **every** cycle; it means with the needle of a front and a back bed. Its yarns are used for connecting face A and face B.

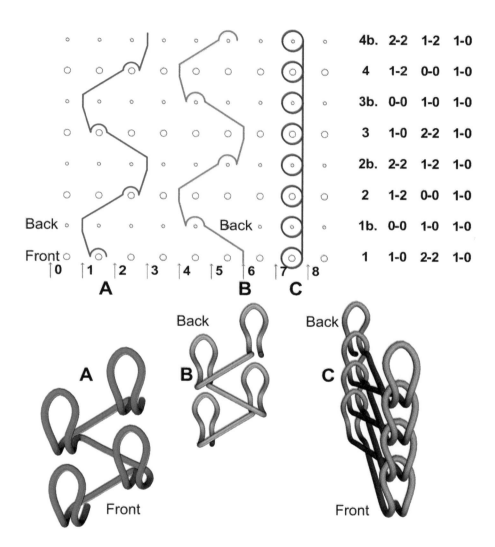

										4b.	2-2	1-2	1-0
										4	1-2	0-0	1-0
										3b.	0-0	1-0	1-0
										3	1-0	2-2	1-0
										2b.	2-2	1-2	1-0
										2	1-2	0-0	1-0
										1b.	0-0	1-0	1-0
										1	1-0	2-2	1-0

FIGURE 10.3 Double face fabrics of three guide bars. Bar A produces the front of the fabrics, bar B the back face and the bar C makes loops on both needle bars and connects the two faces. The figure is for left side patterning device.

10.5   Loop types

The careful reader should have noticed that the selection of the miss-lapping position for passing trough the needles of the opposite needle bed can be different than the selected one in the Figure 10.2. For instance, guide A has in the first cycle motion 1-0/, and stays the for second at 0-0/, but it could move in the direction of the next lap a little bit and pass through the back needle bars with 1-1/ (miss-)lapping movement. Actually, if the position for the second cycle changes to 1-1/ the loops have to be drawn as closed loops, and currently they are drawn as open loops. Careful analysis of the 3D images in Figure 10.2 would give the result that all the loops of the bars A and B are closed loops, but the lapping diagram (2D) shows them as open loops. The explanation of this dissonance can be explained as follows - *the lapping on the opposite needle bar influences the loop type of the current needle bar only, if the yarn is held somehow within the structure of the opposite needle bar.* The yarn can build a loop at the opposite fabric or it can be held between the legs as laid-in - then it is held there and influences the position and orientation of the loop legs at the front needle bar. If only miss-lapping is done at the opposite needle bar, as shown in Figure 10.4, then these miss-lappings do not influence the real type of the loops at the current (in this case front) needle bar.

In this meaning, the lapping Figure 10.4a) and Figure 10.4b) leads to both the closed tricot loop, and are for the loop type - for the crossing of the loop legs equivalent to the closed tricot lap, presented with pink dashed line at Figure 10.4c). The lapping at Figure 10.4d) leads to open tricot loops on the front bar. The miss-lapping positions have to be selected so that the guides have as few additional motion steps as possible and as few accelerations as possible. Good selection of these positions takes care on the yarn loading and on the machine dynamics. Not all positions can be selected for miss-lapping, because at some of them, the yarn will interlace with yarns of other guide bars, depending on their motion. Summarising the current section, for the reader should be clear, that the loop type from the lapping diagram does not always correspond to the real loop type, as this is for single face fabrics. In double face fabrics additionally a third loop type can be found - this of the connecting loops (Figure 10.5). The connecting loops have their legs oriented perpendicular to the plane of the loop and it is not essential for the fabrics, if these are crossing or not. More important is their existence - as these loops build the connections between the two faces in most fabrics and they can be used for holding laid-in yarns. There are rules (see Chapter 7.3.2) for single face fabrics that can be used for determining if a laid-in yarn is held or not by some loop, and these rules are based on the lapping directions of the two bars. Normally the inlay is fixed by counter lapping, and it slides to another position for and equal lapping of the two guide bars. These rules can be

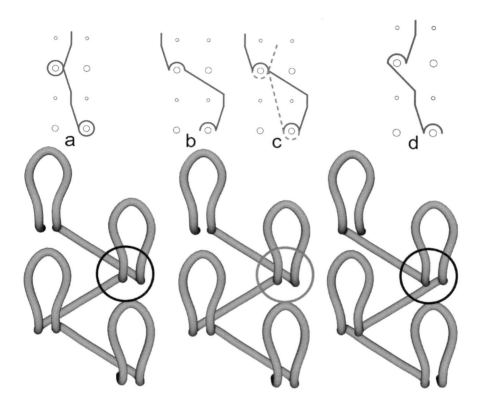

FIGURE 10.4 The lapping diagram for double needle bar fabrics do not represent the closed and open loop types consistently. a) closed tricot b) closed tricot, too, but due to different lapping position on the back needle bar drawn as open tricot c) equivalent structure for a) and b) , d) open tricot on spacer structure.

applied **partially** as well for double needle bar fabrics, taking into account additionally the existence or not of connecting loops.

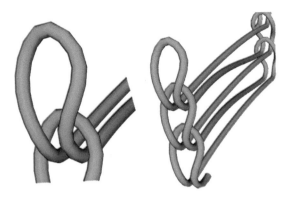

FIGURE 10.5 Connecting loop. The loop legs are going to the next needle bar.

10.6   Double face structures with one guide bar

10.6.1   Chaining laps

The simplest double face structures are produced by one guide bar only. All classical stitch types with repeat of two courses - lock-stitch, tricot, cord stitch, satin, produce only separated chains of loops. Each warp yarn knits on the same pair of needles and there is no connection between these. Figure 10.6b) presents simulated such structure based on the lapping diagram of Figure 10.6a). Such single chains are used as a basic element for several mesh structures. Tricot stitches Figure 10.6c) and cord stitches (Figure 10.7a) have larger distances between the loops of front and back sides (Figure 10.6e and Figure 10.7d))but the chains are still vertical and not connected, as seen at the side view of Figure 10.6f and the top view of Figure 10.7c.

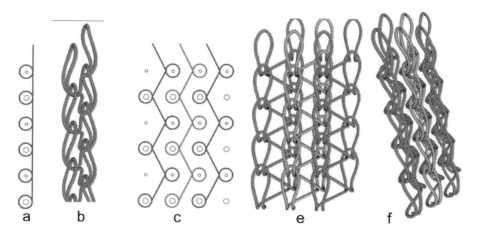

FIGURE 10.6 Chains based on lock stitch and tricot stitch a) lapping diagram, b) chain as 3D simulatoin, c) lapping diagram of tricot stitch e) front view of simulated chains, f) side view for demonstration of the missing connection between the chains.

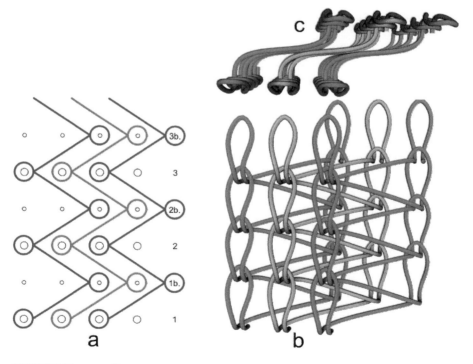

FIGURE 10.7 Chains based on cord stitch a) lapping diagram, b) front view c) top view.

10.6.2   Atlas laps

Two course Atlas lapping movement is the simplest stitch, which produces solid (or spacer) warp knitted structures. The lapping diagram (Figure 10.8a) can be decomposed into two lappings - one for the front needle bar (Figure 10.8c) and one for the back needle bar (Figure 10.8b). The yarns, placed on the needles of the front bar will produce connected structure based on cord stitch (Figure 10.8f) with common inclinations of the loops. If the yarns become placed only on the back needle bed, they would build chains (Figure 10.8e). The connection of one chain and the cord stitch loops build the basic double-needle-bar element (Figure 10.8d), which consists of two alternating loops on the front side and consecutive loops at one and the same needle on the back side. With a guide bar with full threading (Figure 10.9a) this pattern produces solid structure where the single elements are connected in a similar way as interlock weft knitted structures. This interlacement can be recognized by following the colour yarns on the top view of the fabrics (Figure 10.9d) or of its bottom view. Figure 10.10 demonstrates a five-row Atlas lap structure, produced from one full threaded guide bar. Careful analysis of the pattern will show that both sides will have slightly different optics, because of the different distribution of the loop positions.

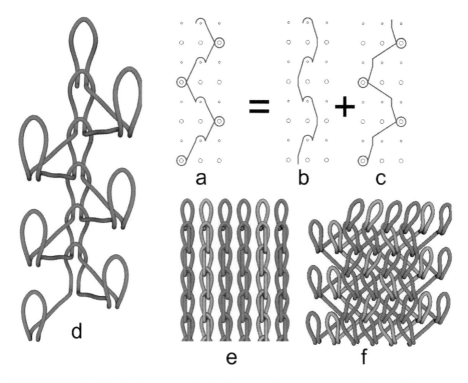

FIGURE 10.8 Atlas lap as a basic element for spacer structures a) lapping diagram, b) effective lock stitch at the back needle bar c) effective cord stitch over the needles of the front needle bar, e) effective back side of straight vales, f) effective front side of with common for cord stitch zig-zag wales. The complete fabric is shown in Figure 10.9.

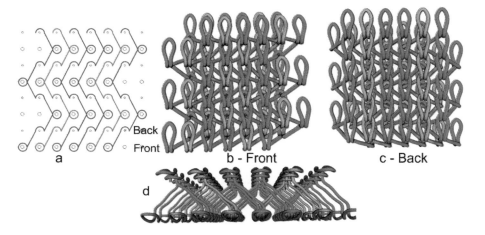

FIGURE 10.9 Atlas lap with full threading a) lapping diagram, b) front view (idealized, without inclinations of the loops)c) back view d) top view.

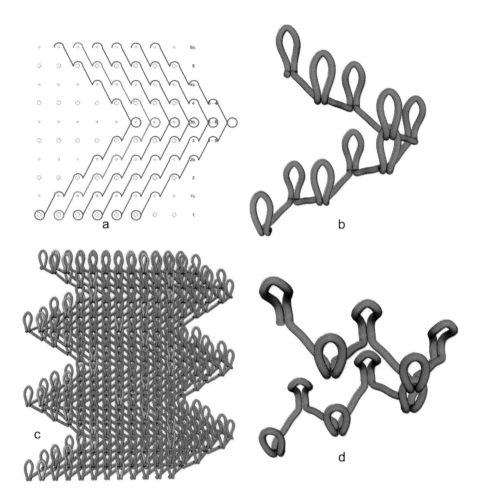

FIGURE 10.10 Five-row atlas with full threading on double needle bar machine a) lapping diagram, b) loops build alternating on the front and back needle bars, based on a single yarn, c) front face of the fabrics d) upper side view of b).

10.6.3   Laps for equivalent both sides

he lapping motion at the front needle bar can be repeated in the back needle bar. Each yarn will build the same loops on both guide bars in this case and the fabrics will have identical appearances. Figure 10.11a shows such one lapping, based on a combination between tricot and lap stitch. Each yarn builds loops over two needles on both needle beds (Figure 10.11b) and these connect with the next yarn. The type of the connection between the loops is visible in the top view in Figure 10.11c. Each kind of lapping can be "doubled" in this way for double needle bar structures. Another example of such sample is presented in Figure 10.12.

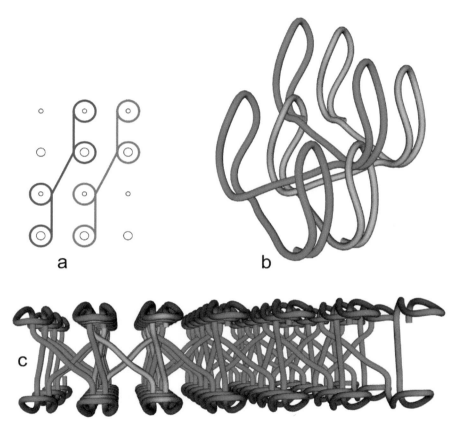

FIGURE 10.11 Tricot with repeating loops on both bars for creating identical faces on both sides.

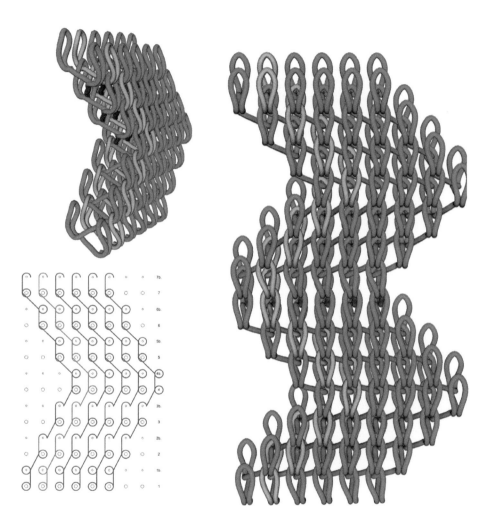

FIGURE 10.12 Atlas lapping with repeated loops on both sides for getting an identical optical appearance.

10.7 Conclusions

All principles for creation of lapping diagrams and building loops of single needle bed machines can be applied for the separated beds of the double needle bar machines. The independent analysis of the loops on two front and back beds allows the decomposition of the pattern and showing an idea about the distributions of the loops on these sides. Some types of patterns for double face fabrics, based of one guide bar, were demonstrated. Although these samples allow some patterning using different lapping or threading, significantly more possibilities are given if multiple guide bars are used, as described in the next chapter.

11

Double face fabrics with multiple guide bars

11.1 Introduction

The real advantages of double face fabrics is the possibility for combinations of different patterns and different materials in one sample. Except the front and back faces, there is a third "face" - the middle area in the spacer fabrics, which is remains hidden for the end user, but is very important for the engineers. The middle area determines the thickness and compressibility of the spacer fabrics, allowing air circulation or it can adopt thicker, cheaper or more functional yarns and wires. This chapter presents different possibilities for connection and patterning of the two faces for both solid and spacer multi-bar fabrics.

11.2 Connection types

Fabrics presented in Chapter 10 have two faces, but are built from **one** yarn system. For this reason they present one solid structure (or single chains) as visualised in Figure 11.1 a1 and Figure 11.1 a2. Using two groups of guide bars with a suitable pattern, but so that the group of the guide bars, working on the front needle bed do not build loops on the back and vice versa, allowing creation of two independent single face structures (Figure 11.1 b1 and b2). Certain lapping motions for selected guide bar arrangements can cause interlacement of these two independent fabrics based on their underlaps as visualised in Figure 11.1 c1 and c2 (such sample is given in Figure 11.11 later). These form configurations that are possible with at least two guide bars on the same machine.

Getting the third guide bar making loops on **both** needle bars allows a large variety of connections between the faces. A full or mostly full threaded connection guide bar leads to a spacer or solid structure with a large number of connecting yarns (Figure 11.2a1 and a2). One needle of it can be enough to connect the two independent fabrics on one side and allow production of a structure with double larger width as the machine width

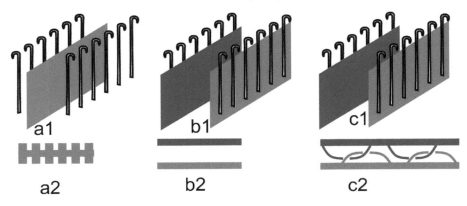

FIGURE 11.1 Basic structures on two needle bars a1) and a2) double face solid fabrics; b1) and b2) two not connected fabrics; c1) and c2) two connected by their underlaps fabrics.

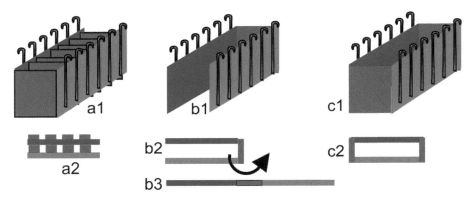

FIGURE 11.2 Connecting faces by third guide bar a1) and a2) regular connection to almost solid or dense spacer structure; b1) and b2) one side connection used for sample width.

(Figure 11.2b1 and b2). Connecting the front and back fabrics on both sides (or multiple) is the way to create tubes (Figure 11.2c1 and c2).

In the case of almost parallel connecting yarns (Figure 11.3a and b) one larger distance between the needle bars leads to spacer fabrics. The space can be increased if the loops on the back needle bar are built few wales by side than the loops of the front bed (Figure 11.3c1). Such fabrics, after removed from the machine can be fixed to a state (Figure 11.3c2). Such fabrics need low shear resistance in order to be able to be deformed to the higher thickness, but this low shear resistance can be as well a problem during use later. Normally for increasing the shear stability the laps are placed so, that builds stable triangles and not a mechanism of four linkages (Figure 11.3d and e). A combination of the pattern of Figure 11.3 b and Figure 11.3 c1 on the same structure can be used for production of spacer fabrics with different thicknesses, as developed and patented by Helbig et al. [7]. The methods of the modelling of such structures were reported in [22] and [12].

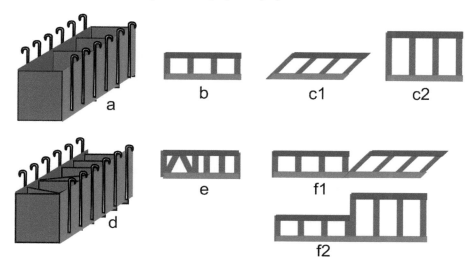

FIGURE 11.3 Backlapped GB: 0-1/3-2/5-4/2-3//.

11.3 Fabrics with two guide bars

11.3.1 Symmetric lapping

One guide bar with cord (or tricot or satin) stitch produces unconnected chains on a double needle bed machine, but adding a second guide bar and using both bars with fill threading changes the yarn interlacing and leads to a complete fabric. Figure 11.4 a shows the lapping diagram of single yarns and Figure 11.4 b their interlacement. At this stage this is still a vertical chain of four loops. If both guide bars become fully threaded, as shown in Figure 11.5, then all needles (except the most outer courses) will be overlapped by yarns of **both** guide bars and will build plated loops. In these loops the yarns of the front and back guide bars are knitted together into one wider fabric. The plating process and the arrangement of the loops is a complex process, described in Chapter 4. In the normal case the front guide bar GB1 will plate

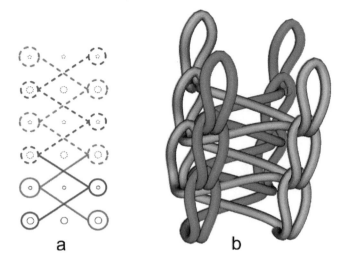

FIGURE 11.4 Cord stitch chain of two guide bars as basic structural element for the double face fabric of Figure 11.5. a) lapping diagram, b) simulated 3d view.

over the loops of back guide bar on the front face. The yarns of the back guide bar (GB2 in this case) plate over the yarns of the front bar on the back face of the fabrics. This alternating process can be used for creation of colour or functional effects with suitable threading and pattern. An identical sample based on tricot stitch, very commonly used for spacer fabrics, is visualized in Figure 11.6. If a half threading is applied for the tricot stitch sample (Figure 11.7) a rib type structure is created. In this case there are no more crossing underlaps and the connection between the guide bars is only based on the common plated loops. This structure does not imitate interlock fabrics, as this is the case for full threading; it looks like rib 1:1 in a double face weft knitted structure, but with vertical columns with the same colour (Figure 11.7d and e).

Atlas type lapping can be used for knitting of simple double face structures with two guide bars, too (Figure 11.8). The loops on each face are build alternating the one and second guides.

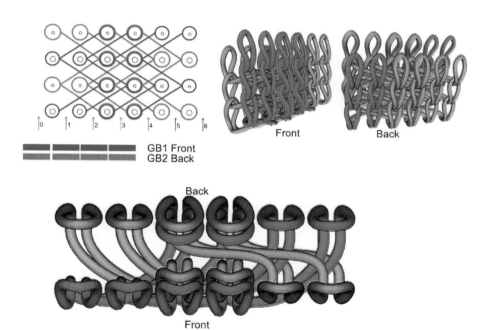

GB1 Front
GB2 Back

Front   Back

Back

Front

FIGURE 11.5 Double face fabrics based on cord stitch, with full threading of the guide bars. The top view illustrates the plated loops well.

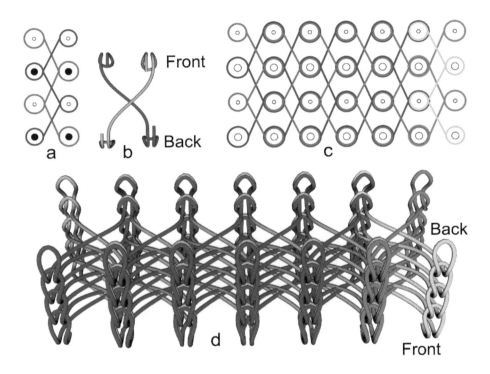

FIGURE 11.6 Double face fabrics based on tricot stitch, with full threading of the guide bars. a) lapping diagram b) crossing of single yarn per guide bar, c) lapping diagram for full threading d) fabrics.

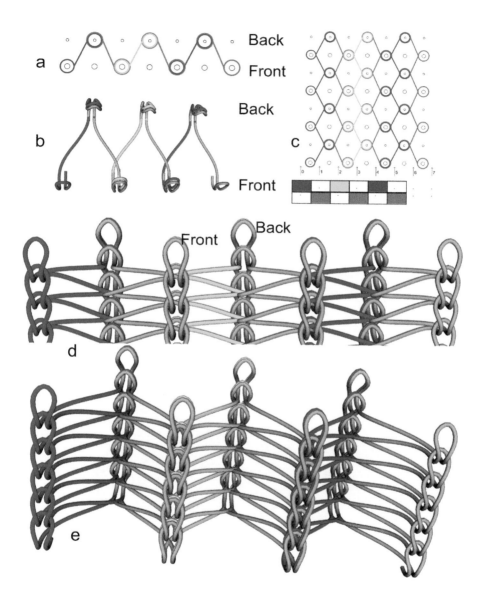

FIGURE 11.7 Double face fabrics based on tricot stitch, with half threading of the guide bars. a) lapping diagram for the first course b) top view of one double course, c) lapping diagram for half threading d) fabrics at small needle bed distance e) fabrics as spacer fabrics for larger distance between the needle beds.

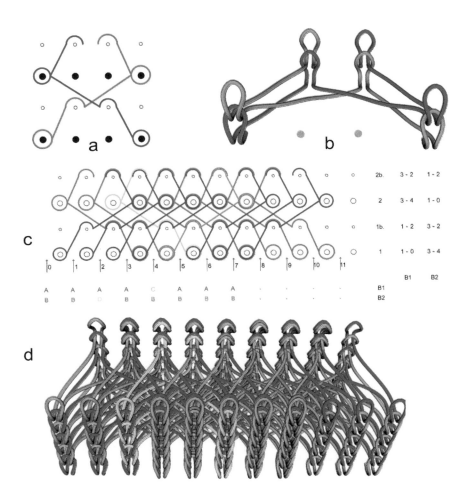

FIGURE 11.8 Double face fabrics based on atlas stitch with two guide bars. a) lapping motion b) basic structural element, c) lapping diagram, d) 3D simulation.

11.3.2 Asymmetric lapping

The tricot stitch of one guide bar with full treading builds vertical chains (Figure 10.6). These chains can be connected together into a fabric using a second guide bar with connecting lock stitch lapping, as visualised in Figure 11.9a. The complete fabrics has asymmetric ribs (11.9a and d), which can becomes better or less clearly visible depending on the run-in settings of the separated guide bars ([28], p. 174). A combination of such fabrics and vertical double-needle bar lock-stitch chains can be used for production of ribbed scarves (Figure 11.10.)

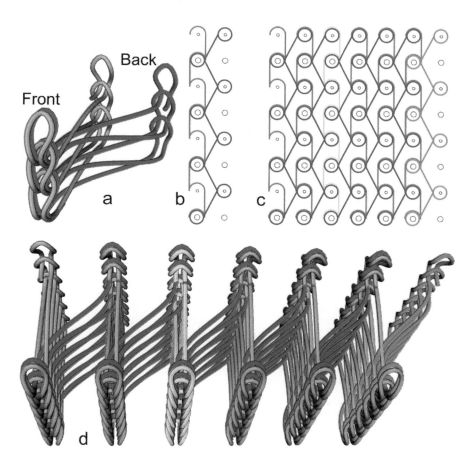

FIGURE 11.9 Tricot-lock-stitch spacer fabrics with asymmetric ribs. a) Connection between single yarns of the both guide bars, b) basic lapping c) lapping diagram with full threading, d) 3D view of the fabrics.

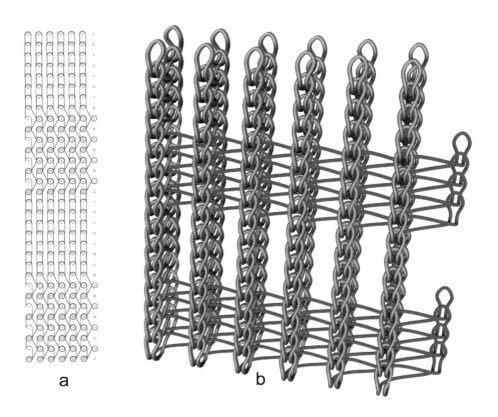

a   b

FIGURE 11.10 Ribbed scarves based on combination of the pattern of Figure 11.9 and lock-stitch based wale chains.

11.3.3 Underlaps connected fabric

As already presented in Chapter 10.3, Figure 10.3, if the front guide bar builds loops on the front needle bed and the back (GB2) guide bar builds loops on the back needle bed, two separate fabrics will be produced (Figure 11.11 a, c, e). Interesting is the configuration if guide bars start not more playing with their one's needle bar and get in love with the opposite one. Like in the real life, the connection between opposite genders (working guide bars) can lead in the warp knitting to love and a more advanced offspring, too. If the front guide bar build loops only on the back needle bar and the back guide bar build loops only on the front needle bar, then the two fabrics (can) get connected with their underlaps (Figure 11.11 b,d,f). The yarns interlace during the motion of the guide bars (Figure 11.11b) and build a soft hidden connection between the two faces (Figure 11.11f). Of course, this interlacement is possible only if the lapping motion of the guide bars causes this, so during the design of such fabrics this has to be carefully investigated. As the connection between the fabrics is based on naturally placed underlaps, the fabrics of this type are more flexible and **softer**, than the double face structures, where the yarns build loops in **both** faces. If the connecting yarn builds loops on both faces, the connections are almost perpendicular to the faces and this increases the compression stiffness. The connection at the underlap level allows small relative motions between the two fabrics and in this way makes the fabrics softer to bending, and the compressibility depends only on the lateral elasticity of the yarns.

11.3.4 Mesh structures

All net (or mesh) structures, created with single needle bed machines, can be extended to their double needle bed equivalents. Figure 11.12a represents the lapping of the guide bars and Figure 11.12b the resulting structure, based on a combination of lock stitch chains and tricot for their connection. Figure 11.12 c) and e) demonstrates the resulting structure for the lapping diagram Figure 11.12d, where half threading of each of the bars is applied.

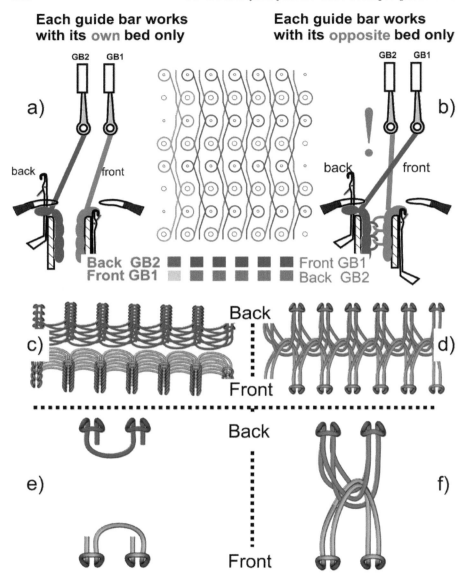

FIGURE 11.11 Front and back needle bar fabrics with the same pattern can be unconnected or connected by the underlaps, if the guide bars switch their pattern. a) each guide bar works with its own needle bar, c) two not connected fabrics are built, e) top view over single yarns of the both guide bars; The two fabrics can be connected by the underlaps if the yarn crosses: b) each guide bar works with the opposite needle bar, d) view of the connected fabrics, f) top view of the connected single yarns of the both guide bars.

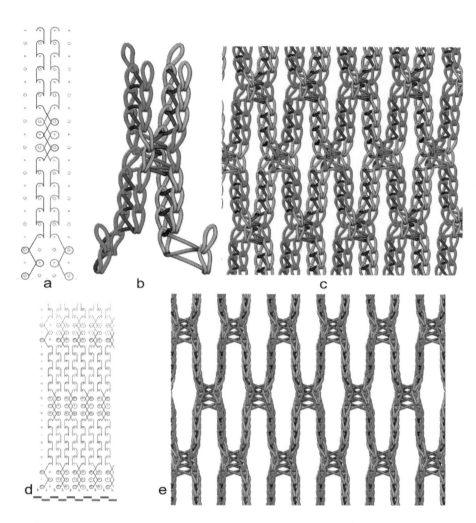

FIGURE 11.12 Mesh type double needle bar structure a) lapping motion of the two guide bars, b) 3D side view of the single yarn loops of the both guide bars, c) element of the structure d) lapping diagram with threading for the samples c) and e) front view.

11.3.5   Laid in double face fabrics

The rules for connecting of laid-in yarns in the loops for single needle bed structures were presented in Chapter 7. These rules remain valid for double needle bar machines too, which means that only the yarns, placed with laid-in lapping from guide bars located **between** another guide bar(s) and the corresponding needle bar during the underlapping step, will be connected between the loops and the underlaps of these loops. In the case of double needle bar machines, the numbering of the guide bars, according DIN ISO 10223:2005, has to be taken into account. The rule from single face fabrics has to be applied with understanding the guide bar orientation. For the double needle bed machine the proper arrangement of the guide bar is demonstrated in Figure 11.13. The laid-in yarn of guide bar GB2 (or every guide bar, that is between the needles and the loop building guide bars during the underlapping motion at the back needle bed) will place laid-in yarns, which are connected between the the back face loops and their underlaps or the front bed loops (Figure 11.13a). The laid-in yarns of the first guide bar (GB1) will be connected between the loops of the front needles and their underlaps (or the loops of the back needle bed), if there is a guide bar after that, like GB2 in the case of (Figure 11.13b). Normally, one guide bar produces only loops from finer and stronger yarns or place only laid-in elements from cheaper and thicker yarns. In one such "standard case", only the laid-in motions at one of the cycles will lead to fixation of the yarn. Figure 11.14 demonstrates a lapping diagram and a product, where each second cycle the red yarns are placed horizontally between the loops and the other half cycle the guide is doing miss-lapping. In this way thicker, voluminous yarns can be placed in order to increase the thermal insulation of the fabrics, while thinner and more expensive yarns are used for creation of the two faces with loops.

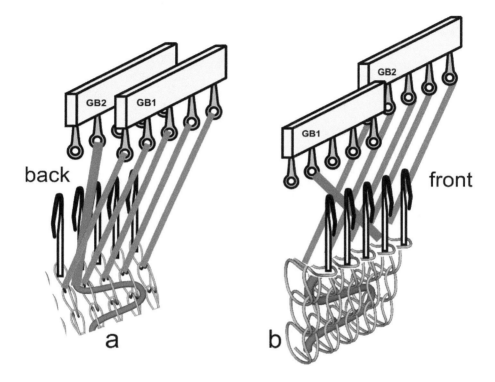

FIGURE 11.13 Configuration of the guide bars for laid-in placement on double needle bar machines a) the laid in yarn can be connected in the loops of the back needle bar, if there is a guide building loops before it, b) the laid-in yarns can be connected between the loops and underlaps on the fabric on the front bed, if the laid-in yarn is placed by the first guide bar.

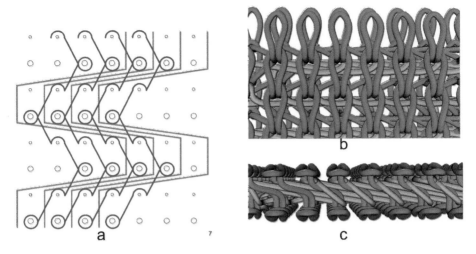

FIGURE 11.14 Laid-in yarns between the two faces on double needle bed structure. a) lapping diagram, b) front view, c) top view.

11.4 Fabrics with three and more guide bars

11.4.1 Spacer fabrics with four or six guide bars and laid-in

The idea of the fabrics from Section11.3.5, Figure 11.13 can be applied for production of two fabrics, which can be connected together by one or two lock stitch guide bars into one spacer fabrics, as reported in [8] with two connecting guide bars or in [28] with only one connecting bar. The lapping movement of such case is demonstrated in Figure 11.15. This is a simple and most common construction for spacer fabric or for sport shoe fabric. For spacer fabric the connecting yarns of guide bars b) are threaded with stiff monofilaments in order to provide stability of the connecting yarns. For sport articles the connection is soft multifilament yarn.

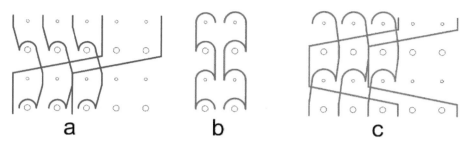

FIGURE 11.15 Lapping diagram for six guide bar spacer fabrics. a) GB1 and GB2 build the front fabrics based on lock-stitch and laid-in, Lapping movement GB 1: 0-0/0-0/3-3/3-3// ; GB 2: 0-1/1-1/1-0/0-0// ; b) guide bars GB3 and GB4 are making connection between the two layers GB 3: 1-0/0-1/1-0/0-1// ; GB 4: 0-1/1-0/0-1/1-0// ; c) GB5 and GB6 are making the back face fabrics again as lock-stitch and laid-in GB 5: 0-0/0-1/1-1/1-0// ; GB 6: 3-3/0-0/0-0/3-3//.

11.4.2 Spacer mesh fabrics

Six guide bar mesh structure [28] is demonstrated in Figure 11.16. In this case the same principle as in the previous sample is applied - two guides are building mesh on the front (Figure 11.16a and b) only, another two the mesh on the back side (Figure 11.16e and f) and the connecting part in this case is done by the two middle bars in tricot lapping in opposite motion (Figure 11.16c and d). Note, that the front and back structures require half threading in order to get the mesh effect.

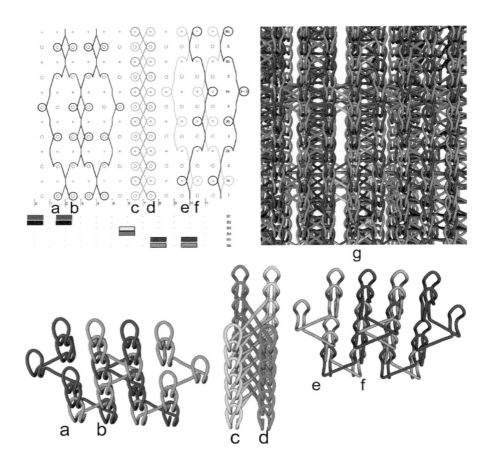

FIGURE 11.16 Six guide bar spacer mesh structure, a) and b) - build the front fabrics, e) and f) the back fabrics, c) and d) the connection between the layers. g) presents the complete structure.

11.4.3 Samples

Figure 11.17 presents photos of a high-distance warp knitted fabric with two different faces. The one face, Figure 11.17 A, has mesh type pattern to ensure better air permeability, while the other face, Figure 11.17 B, is designed as a complete flat surface. The monofilament yarns are connecting the two faces and keep the higher distance in this case of 12mm between Figure 11.17 C. They build loops on the both faces plating the softer multifilament yarns of each face.

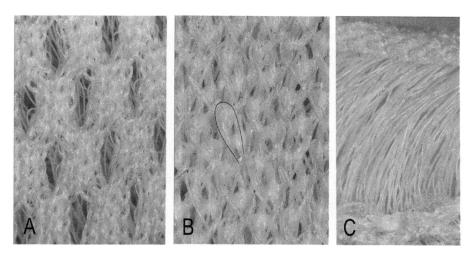

FIGURE 11.17 Spacer fabric with two different faces A) mesh type B) normal plane structure, C) connecting monofilaments yarn.

Such fabrics with a thickness between 3 and 5 mm are used for improving wear comfort at the contacting area between the body and rucksack and in shoes. The mesh type structure has its natural gaps, which allow good air circulation at the contact area, and the filaments have less density, so that air circulation within the spacer fabrics is possible too. Microscopic image of such fabrics is given in Figure 11.18.

FIGURE 11.18 Mesh type spacer structure with lower pile density for better air circulation.

11.4.4 Hollow fabrics

As explained in Figure 11.3 if only one or two sides of the two single needle bar fabrics are connected by single yarn, tubular or wider structures can be obtained. The wider structures will be not discussed here, because they can be produced by removing the connection on the one side. Figure 11.19 demonstrates a simple way for building tubular fabrics, where both connections are done with one and the same guide bar. In this case a tricot stitch is used, and its yarns are marked blue. This structure will have two loops per repeat, which are plated - and build both of yarns of the guide bar and yarn of the connecting yarn. If such connection is not wished because it influences the mechanical properties, the pore size or the aesthetics of the fabrics, then two separate guide bars for the left and right hand side connections have to be used. Such clean connection is demonstrated in Figure 11.20. For production of bifurcations, for instance, of medical products or of panty hose, two additional connection places have to be placed in the middle of the machine, in order to build two separate connected tubes.

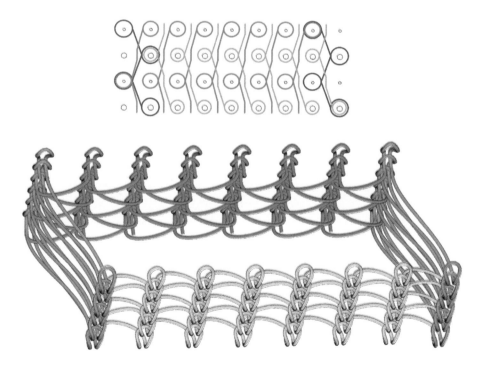

FIGURE 11.19 Tubular spacer fabrics with simple connection, realized by one additional guide bar.

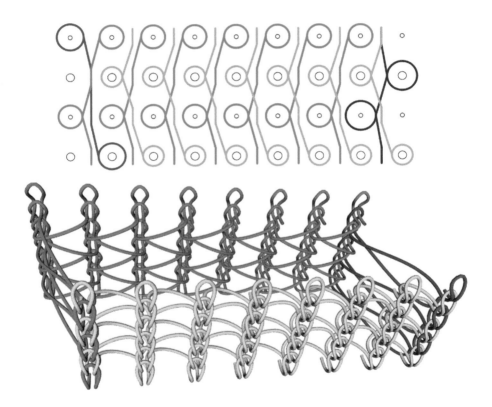

FIGURE 11.20 Tubular spacer fabrics with clean connection, built of two separated guide bars for each side connection.

11.5   Jacquard

The most advanced warp knitting machine is the double needle bar Raschel with jacquard guide bars. Common configuration of the guide bars for the RDJ and RDPJ machines of the Karl Mayer company [1] is demonstrated in Figure 11.21. The piezo-jacquard bars are positioned in the middle of the machine and can work with both needle bars. The first bar(s) does not work with the back needle bar and the last bar(s) does not work with the front needle bed, which allows optimization of the space and the dynamics of the machine. Using jacquard bars allows production of fabrics with different colour effects, structures, connections and becomes widely used for production of fabrics for shoes, or pre-made components of clothing and textiles during the actual knitting process. The development of the jacquard fabrics on double needle bar machines requires deep understanding of all types of patterns and rules in this book, because the double needle bed jacquard structures are combinations of different elements of these at different places and needle beds.

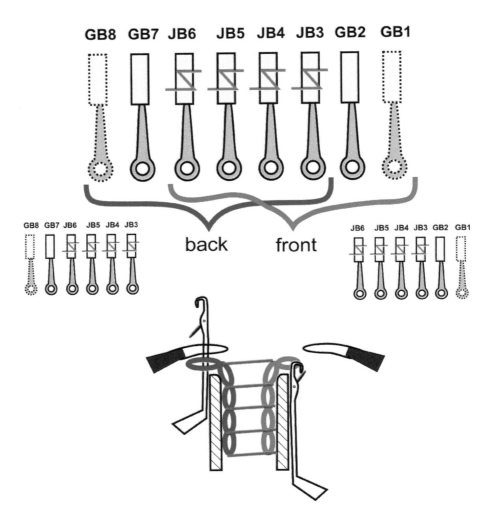

FIGURE 11.21 Arrangement of the guide bars on double needle bar jacquard machine.

Part V

Engineering design

12

Calculations

12.1 Introduction

The engineering design of warp knitted fabrics today requires the estimation of their main properties, where at the first place this is the **specific weight** of the fabrics. For the calculation of the weight, the yarn length in an unit cell has to be known. The yarn length is required as well for the preparation of beams, the production process and for the adjustment of the machine. This chapter presents some models and equations for the determination of these parameters.

12.2 Yarn length per loop

12.2.1 Lenght of the loop head

The yarn length for one loop and the following underlap can be determined geometrically using a model, like the presented one in Figure 12.1. The loop is divided into two elements - loop **head** and loop **arms,** and the transition curves between the arms and the underlap are ignored. Assuming, that the loop head is a half circle arc (Figure 12.2) with radius r, the yarn length in the head will be

$$L_{head} = \pi \cdot r_{head} \tag{12.1}$$

where r_{head} is the radius of the curvature of the head. This radius can be determined in different ways. During the loop building process, this radius depends on the needle shaft diameter d_{needle} and the yarn diameter d_{yarn}, and the equation can written as:

$$L_{head} = \pi \cdot r_{head} = \frac{1}{2} \cdot \pi \cdot (d_{needle} + d_{yarn}) \tag{12.2}$$

After the knitting process, the loop head can relax and change its form, depending on the bending stiffness of the yarns and the number of the yarn

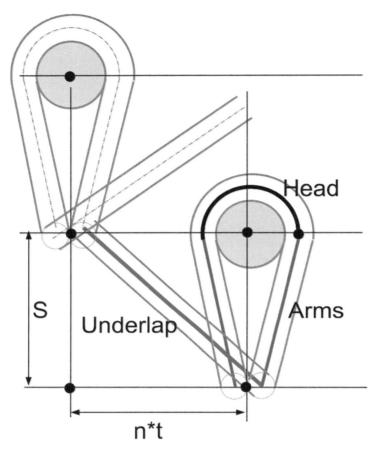

FIGURE 12.1 Geometrical model of warp knitted loop for the yarn length calculation.

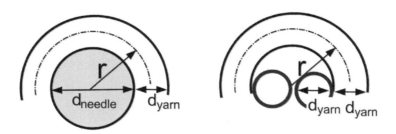

FIGURE 12.2 Models for the calculation of the loop head length.

arms going through it. In the case of single guide bar of yarns with the same diameter, the loop head nominal radius can be obained as $r_{head} = d_{yarn} + 0.5 \cdot d_{yarn}$, so that the complete loop head arc has a length of

$$L_{head} = \pi \cdot \left(d_{yarn} + \frac{d_{yarn}}{2} \right) = \pi \cdot \frac{3}{2} \cdot d_{yarn} = \pi \cdot 1.5 \cdot d_{yarn} \qquad (12.3)$$

In the book "Warp knitting production" Dr. S. Raz [19] whose paper "Warp knitting Calculation Made Easy" [3] is cited, recommends the calculation of the loop head as an circular arc with radius $2 \cdot d_{yarn}$:

$$S = \pi \cdot 2 \cdot d_{yarn} \qquad (12.4)$$

According to the documentation of the Karl Mayer company [2] the loop head can be calculated using empiric equation using the needle diameter:

$$L_{head} = \frac{1}{2.2} \cdot \pi \cdot d_{needle} \qquad (12.5)$$

These equations all are related mainly to (flexible) textile multifilament yarns. For monofilament yarns and metal wire the loops heads have larger radius, because of the larger bending stiffness of the material (Figure 12.3). In this case the loop head radius should be measured from the image, as it is significantly larger than the diameter of the needle and the diameter of the wire.

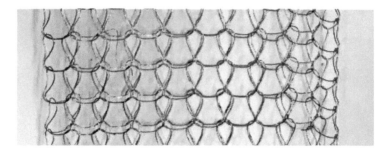

FIGURE 12.3 Metal loops of (weft) knitted metal wire [10] have a larger radius of curvature because of the bending stiffness of the material.

Depending on the type of the structures, properties of the yarn material and the available information, the reader can choose which of the equations 12.2, 12.3 or 12.5 is the most suitable one for the current situation. The equation 12.5 is a good choice for standard structures and if the machine needle diameter is known.

12.2.2   Length of the loop arms

The length of each loop arm can be considered as equal to the height of the course S (Figure 12.1), which can be derived from the number of the courses per cm (CPC) or the number of the courses per inch (CPI):

$$L_{Arm} = S = \frac{10}{CPC} = \frac{25.4}{CPI} \tag{12.6}$$

The equation 12.6 actually ignores the inclination of the arms from the vertical direction (line segment AC in Figure 12.4), based on the larger width of the loop in the upper side. Through the loop head the needle goes and later the arms of the next loops. The consideration of this angle makes the arm length calculation more exact, but as well the equation more complicated for the normal workers, as it has to use root function after the application of the Pythagoras theorem for the triangle $O_0 O_1 D$ or ABC. Since $O_0 O_1 = BC = \frac{d_{needle}}{2}$:

$$L_{Arm} = AC = O_0 D = \sqrt{S^2 + \left(\frac{d_{needle}}{2}\right)^2} = \sqrt{S^2 + \frac{d_{needle}^2}{4}} \tag{12.7}$$

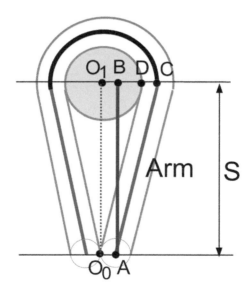

FIGURE 12.4 Geometrical model for determination of arm length.

12.2.3 Underlap length

The length of the underlap can be considered simplified as the distance between the swing-out position of the current cycle and the swing-in position of the next cycle. Using the spacing between the needles, based on the machine gauge E:

$$t = \frac{25.4}{E} \tag{12.8}$$

and the number of the spaces between the two cycles n, the length of the underlap will be:

$$L_{Underlap} = n \cdot t \tag{12.9}$$

Here again, it should be considered, that the exact length of the underlap depends as well on the take-off speed of the machine, and the longer the loops are the less accurate the equation 12.9 will be. Applying again the Pythagoras theorem for the triangle with cathetus $n \cdot t$ and the loop height S, the underlap length can be calculated as:

$$L_{Underlap} = \sqrt{S^2 + (n \cdot t)^2} \tag{12.10}$$

12.2.4 Complete equation

The complete length of the loop and the corresponding underlap is the sum of all parts

$$L_{total} = L_{Head} + 2 \cdot L_{Arm} + L_{Underlap}. \tag{12.11}$$

Using the simple equations from the above parts, the following equation can be derived:

$$L_{total} = \frac{1}{2.2} \cdot \pi \cdot d_{needle} + 2 \cdot S + n \cdot t \tag{12.12}$$

This equation is the simplest way for an approximate estimation of the length of the loop in one cycle and the derivation of its run-in value and the specific weight of the fabrics.

A more precise result can be obtained with the more accurate equations:

$$L = \frac{1}{2.2} \cdot \pi \cdot d_{needle} + 2 \cdot \sqrt{S^2 + \frac{d_{needle}^2}{4}} + \sqrt{S^2 + (n \cdot t)^2} \tag{12.13}$$

The equation 12.14 looks more beautiful and regrettably would cause headaches for even more "engineers" (here the quotation marks are used to demonstrate irony), who are looking for an App or software for its calculation instead of taking the calculator or Excel sheet or any other computational software environment like Matlab, Freemat, Octave, Python, Visual Basic, etc. Indeed, modern CAD software for warp knitting, like the Texmind Warp Knitting Editor [11] has integrated the loop length calculation and is able to provide this per each loop and each guide bar (Figure 12.5):

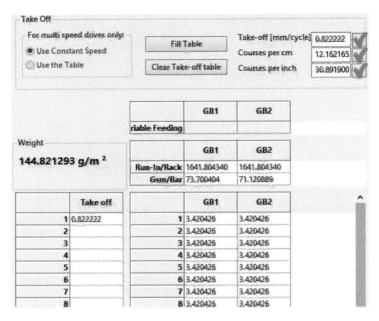

FIGURE 12.5 Example of calculated loop lengths within Texmind Warp Knitting Editor.

12.2.5 Grosberg model

As mentioned in the previous section, the above models do not consider the mechanical properties of the yarns. Prof. G. Grosberg considered the yarn as "elastica" - elastic body which takes its form minimizing its potential energy under the consideration of the external constraints (Figure 12.6 [6]. "From the relations which imply that the length of the elastica is equal to 2.543 times its height" [19], p. 510, the part for the length of the loop head and loop arms is derived:

$$L_{head\,and\,arms} = 2.543 \cdot S \tag{12.14}$$

The underlaps are calculated in the same way as in the previous section. After that, different correction factors are applied in order to consider the loop root parts for the front and back guide bar loops, because at the root these have different lengths. The final equations become for the first and second guide bars [19]:

$$L_{total,GB1} = \sqrt{S^2 + (n \cdot t)^2} + 2.543 \cdot S + 7.12 \cdot d \tag{12.15}$$

$$L_{total,GB2} = \sqrt{S^2 + (n \cdot t)^2} + 2.543 \cdot S + 4.69 \cdot d \tag{12.16}$$

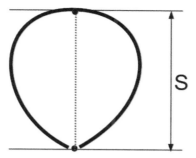

FIGURE 12.6 Form of the loop head and loop arms as elastic body, whose length depends only on the loop height S.

12.2.6 Machine State Loop Model of S. Raz

S. Raz developed in his Ph.D. thesis a machine state loop model[19], p. 511, where the relations for the yarn consumption is based on the state of the fabrics on the machine.

The straight underlap is corrected so that its beginning and end take into account the position and the diameters of the yarns (Figure 12.7). This information was neglected in the previous models, which take the center of

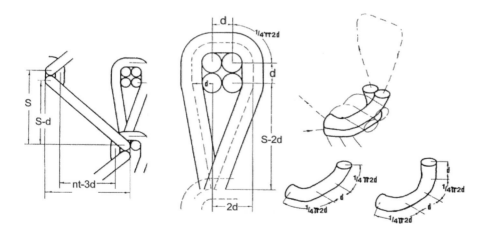

FIGURE 12.7 Machine state loop model for two guide bar fabrics of S. Raz [19], p. 511. In order to unify the equations in this chapter the loop height c from the original is renamed into S, and the distance between the needles renamed to t.

the loop as basis:

$$L_{underlap} = \sqrt{(S - d)^2 + (n \cdot t - 3 \cdot d)^2} \qquad (12.17)$$

The loop head is presented as a set of straight parts and circular arcs with radius equal to one yarn diameter (radius of the inner yarn and radius of the loop yarn):

$$L_{head} = 3 \cdot d + \pi \cdot d \qquad (12.18)$$

The loop arms height is as well corrected in its exact position with one and the other with two yarn diameters:

$$L_{arm} = \sqrt{(S - 2 \cdot d)^2 + d^2} \qquad (12.19)$$

$$L_{arm,other} = \sqrt{(S - 2 \cdot d)^2 + 2 \cdot d^2} \qquad (12.20)$$

The small yarn pieces in the root of the loop are as well considered as straight yarn pieces and as 1/4 of the circular arcs:

$$L_{root} = 2 \cdot \pi \cdot d + 3 \cdot d. \qquad (12.21)$$

Finally the equation becomes the form:

$$L_{total} = \sqrt{(S - d)^2 + (n \cdot t - 3 \cdot d)^2} + \sqrt{(S - 2 \cdot d)^2 + d^2}$$
$$+ \sqrt{(S - 2 \cdot d)^2 + 4 \cdot d^2} + 15.4 \cdot d \qquad (12.22)$$

S. Raz reported that this equation has very good agreement to the practical run-in values.

12.3 Other structural elements

12.3.1 Laid in (Weft) Insertion

The length of the partial weft inserts (laid in) can be assumed to be equal to the loop height S with some addition between 10 and 40% for the arc around the next needle (Figure 12.8):

$$L_{weft} = S \cdot K, \qquad (12.23)$$

where $K = 1.1 \div 1.4$ ([2]).

For placing weft yarn under more needles the simplified equation is based on the needle distances (pitches) and addition of $0.5 \cdot t$ for the reversing point:

$$L_{weft} = (n + 0.5) \cdot t, \qquad (12.24)$$

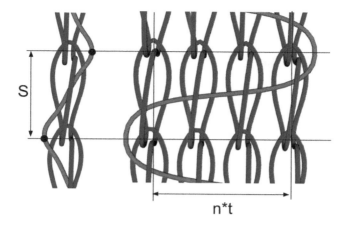

FIGURE 12.8 Laid in yarn around one course and between four courses.

A more correct equation considers both the loop height S and the spacing, analogous to the underlap calculation (Figure 12.9):

$$L_{weft} = \sqrt{S^2 + (n \cdot t)^2} \tag{12.25}$$

Of course, this equation can be made even more precise, if the inlay is considered as an arc around the loop root and a line segment

$$L_{weft} = \sqrt{(S - 2 \cdot d)^2 + (n \cdot t + d)^2} + \pi \cdot d \tag{12.26}$$

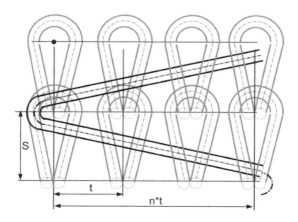

FIGURE 12.9 The exact length of the weft yarn depends as well of the take off.

12.3.2   Loops over two needle

The length of the loops over two needle can be calculated as a sum of the loop lengths of the both single loops and the underlap between these. The underlap length between the two loops is proportional to the needle space.

12.3.3   Double face and spacer fabrics

The double face fabrics and the spacer fabrics are produced on a machine with distance FA between the two needle beds. The yarn of the connecting underlaps can be calculated using this distance and the number of the needles spaces $n \cdot t$ between the both lapping positions: per

$$L_{connection} = \sqrt{FA^2 + n \cdot t^2} \qquad (12.27)$$

12.4   Lengths per rack

In warp knitting one rack is used for the yarn length measurement, which covers 480 stitches. For manual calculation the loop length per each loop within one repeat is calculated and then used for the calculations:

$$L_{repeat} = L_1 + L_2 + ... + L_N = \sum_{i=1}^{N} L_i \qquad (12.28)$$

where N is the repeat length (for the basic stritchs, as tricot, cord, velvet $N = 2$) and then the average run-in per rack is calculated as following:

$$L_{rack} = 480 \cdot \frac{L_{repeat}}{N} \qquad (12.29)$$

Length of the yarn in the repeat and the run-in per rack is calculated for each guide bar, which has different lapping.

12.5   Fabric weight

The theoretical fabrics' weight per square meter can be obtained by the calculation of the mass of all yarns in one repeat, divided to the surface area of this repeat. As repeat width has to considered, the least common multiple of the threading strings of all guide bars, with taking into account the empty guides too. In Figure 12.10 both guide bars have threading 1 in, 1 out, so the repeat is 2 needles, and the width of the repeat is $w = 2 \cdot t$, where the t is the spacing

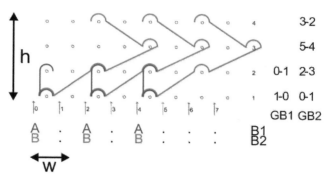

FIGURE 12.10 Lapping diagram of two guide bar samples for calculation of the specific weight.

between the needles. The height of the repeat depends on the pattern and is the least common multiple of the pattern lengths of all guide bars. For the case of Figure 12.10 for open lock stitch this is 2, the case of the back-lapped atlas - 4, the least common multiple is 4. If all loops have the same hight, the repeat height will be then $h = 4 * h_{loop}$. The surface of the fabrics on *the machine* will be $S = w \cdot h = 2 \cdot t \cdot 4 \cdot h_l oop$. The yarn lengths have to be calculated for the completed repeat. For the atlas there are the four loops and underlaps, and the the open lock stitch has to be extended to the repeat and four loops have to be considered there too. The mass of the yarns in the repeat can be calculated using the linear density as $m_A = L_A \cdot T_t$. In this matter, the sum of all yarn masses m_{total} in the repeat divided to the surface of the repeat S will provide the surface length. The units for calculation are not provided here as this task belongs to the first semester in the textile engineering program. The calculations can be done manually, but for a more complicated pattern this does not make sense, as modern software has these equations integrated and is able to provide the data automatically. It has to be taken into account that the calculated values are of the fabrics *on the machine*. After the fabrics are produced, the sample relaxes and changes its dimensions. After relaxation it could/should go through thermofixation process where again other dimensions are fixed and other weight per unit surface will be received.

12.6 Conclusions

The basic geometrical models for the loops allow calculation of the main parameters of the structures such as yarn length, fabric weight, length per rack. These can be calculated based on the machine configuration (machine state) or based on the dimensions of the relaxed loops. All these calculations have to be considered as approximate values, because the relaxation and thermofixation process changes the dimensions of the fabrics. These changes depend on the yarn tension and the fixation parameters and can not be predicted without knowledge of these values and the complete relaxation behaviour of the yarns. All these equations are integrated in modern software and can be accessed automatically there.

13

Software for design of warp knitted structures

13.1 Introduction

This chapter presents some possibilities for the development of warp knitted structures with the software TexMind Warp Knitting Editor 3D. The reader can get an impression about how the main elements of the fabrics definition - such as yarn parameters, machine definition, pattern, take off and 3D image can be defined, calculated and exported for different purposes.

13.2 Historical remarks

Of course, additionally to the TexMind software, there are two or three other software packages available on the market, but the decision for using the Texmind software was very simple - it is developed by the book author and because of this, the author had the freedom to use, adjust and extend it in any way suitable for book publication. The world-wide first **3D** warp knitting simulation system was presented by ALC Computertechnics, mainly known as supplier for the Karl Mayer company during the ITMA 2007 in Munich [20],[21]. This 3D module was developed by the book author in cooperation with and for Dipl.-Ing. Wilfried Renkens, from ALC Computertechnics, Aachen, Germany. After some turbulent times for the company ALC in the difficult for the industrials year 2008-2009, a second generation simulation module Loop3D was developed as a Matlab module, with a very simplified lapping editor. In 2014 TexMind launched a small and simple program, mainly designed for students and for preparation of the simulation data - TexMind Pattern **Painter**. During the years, the **Painter** has grown with functions in order to cover customer requests. It was re-written in 2015 in order to be optimized for a more user friendly interface and to become suitable for machine export, and so the TexMind Warp Knitting Pattern **Editor** was born. The Editor supports machine export for machines of few producers, calculation of run-in and weight of the fabrics. The 3D Version - TexMind Warp Knitting Editor **3D** ()was officially launched as pre-release in 2018, but used in various research and development projects for industrial and academic partners since 2016. This overview covers only some main functionalities, available in the software in 2018. For more detailed information the User's Guide to the current software release has to be used.

13.3   Machine definition

The first step in the development of new pattern is the definition of the machine type and its parameters. Figure 13.1 demonstrates the main settings, from which the most important for the lapping diagrams are:

- Machine Type - single needle bed or double needle bed machine - this selection influences the type of drawing and visualisation of the lapping diagram and as well the behaviour of the software when patterns from the libraries are inserted. For instance, if tricot pattern is inserted in the single needle bed machine, it just comes as normal tricot; inserting tricot on double needle bed machine adds each second cycle misslapping, so that the same tricot can be created on the **one needle** bed only

- Tempi - 2 tempi or 3 tempi - this setting switches on or off the visualisation and saves the positions for the third motion step in some older machines, which run in 3 tempi

- Number of guide bars and their types. Here the number of ground bars, weft bars, jacquard bars can be entered. The software creates the corresponding tables for the motion control of these bars.

For the calculation of the run-in length and the weight of the fabrics, additionally to the above data, the needle head diameter, the machine gauge and the distance between the needle beds (for double needle machine) have to be defined. The numeration of the guide bars from left to right or opposite can be changed "on the fly" any time. The needles on the drawing are a parameter used mainly for convenience when creating images for documentation.

The created machine can be saved as XML file and used at later times, so that the user can create a set of all required machine definition files at the beginning and load these each time when it is required.

Creative designers have different work-flows than the textile engineers - they try to create new samples without limitation of some certain machines and they need to add different guides "on the fly". Adding or deleting guide bars when some pattern already exists is possible, but then the "Init" button has to be pressed, so that all required fields for the new guide bar are generated. Even if in the current version this operation is already arranged to work with possible minimal loss of existing data, such loss cannot always be avoided. For the programmers it possible to predict, list and cover all possible combinations of operations of the user (which type of guide is deleted, when new one is added etc.), but the number of the combinations of these operations is very large and their complete coverage would require significantly larger time, which nobody likes to pay. For this reason, it is recommended for these situations, where it is not clear how many guide bars will be required - with a machine with larger number to be started and not all to be used.

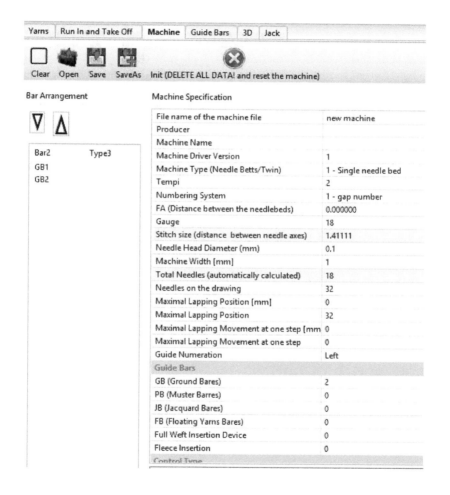

FIGURE 13.1 Options for the definition of the machine configuration.

13.4   Yarn definition

Each yarn type was assigned its own line thickness and colour for the drawing on the lapping diagram (Figure 13.2). If in one guide bar some curves (which represent single yarns) need a different colour or thickness, a separate yarn was to be created for these. For the calculation of the specific weight additionally the yarn fineness in dtex has to be specified. For the usage of the 3D version of the software as well the yarn diameter in mm has to be specified. The software keeps as well data for a smaller yarn diameter for yarns with elliptical cross sections, but these can be visualised at the current time (2018) only with the Loop3D external module; the build in 3D models currently uses only circular cross sections. The remaining material parameters, represented in Figure 13.2 are used currently only from **some** of the exporting functions for FEM computations and from the mechanical simulation module, which is not explained in this book. Each yarn can be saved in a single XML file and reused at later time.

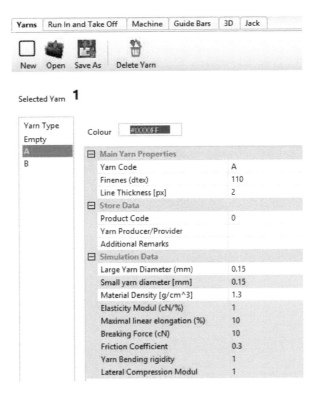

FIGURE 13.2 Definitions of the yarns.

13.5   Pattern editing

Once the machine and the yarn types are defined, the work on the lapping diagram can begin. Figure 13.3 presents the main panel for the pattern visualisation. The lapping of each guide bar can be made visible or hidden with a corresponding tick of the visibility box. There are different ways of editing the pattern. For beginning students, it could make sense to use the "click at swing" method, where for each "swing-in" and "swing-out" motion one mouse click in the corresponding needle gap has to be performed. In the list box of the "Editing" panel (Figure 13.3 down left), the guide bar has to be selected, which has to be edited in the current time. In this case the yarn of the **first guide** becomes visualised with black colour, independent of whether this guide has yarn in it or not (Figure 13.4). For better visualisation, lapping movement can be extended to any number of vertical and horizontal repeats. For the pattern, where each guide has a different repeat, the extension to the full pattern repeat gives the impression of the smallest "unit cell" of this pattern.

Each mouse click in "edit-modus" on the panel is registered, and if it presents valid information is transferred to a number for the numerical coding of the guide bar motion (Figure13.6). Each column there can be selected by clicking its label and after a click to the single pattern from the pattern library (Figure 13.5) assigns this pattern to the corresponding guide bar. The lapping in the selected column can be moved up and down through the knitting cycles, left or right (reducing or increasing the numbers) or mirrored vertically or horizontally with the suitable buttons from the "Edit Menu" (Figure 13.6). Automatic changes of all loops from open to closed and vice versa can be done as well with the "open <> close" button.

The lapping motion of each guide bar can be saved as is or loaded from a separate, human readable text file, through the menu, coming after the right mouse bottom is clicked over the column label (Figure 13.7). The first menu allows as well the editing of the lapping motion on a single text line, which is more convenient in some cases than the going through the cells.

The lapping file is a text file, which can have comment lines, beginning with % and allows empty lines, as demonstrated on the following listing:

```
% Format for Chain Links Rev. 2011-2018 (C) Kyosev
% Comment are allowed after %
% empty lines are allowed
Name:
Lapping Name (optional)
Tempi:
2
Chain notation:
1 - 0 / 1 - 2 / 1 - 0 / 1 - 2 / /
% notation ends by /  empty places are no problem.
```

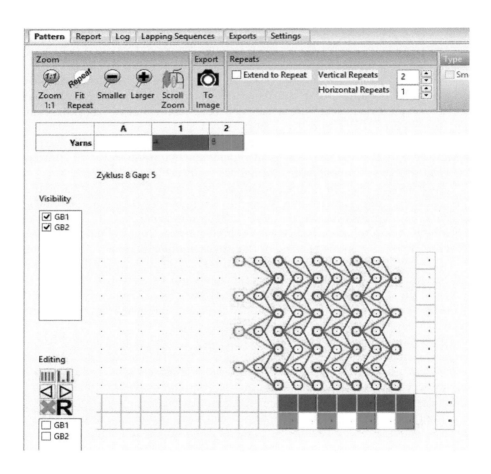

FIGURE 13.3 Pattern definition.

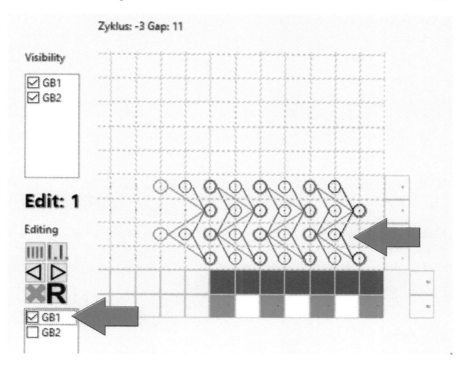

FIGURE 13.4 The selected guide GB1 in editing mode.

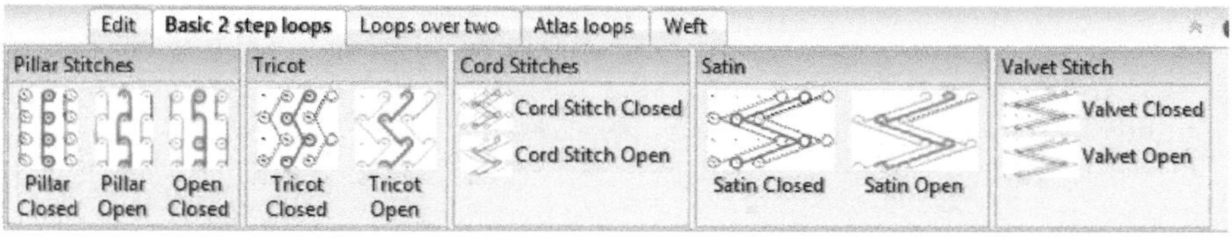

FIGURE 13.5 Pattern library example.

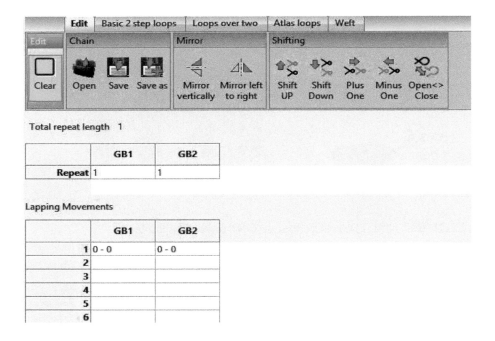

FIGURE 13.6 Lapping editor with possibilities for editing of the numerical input.

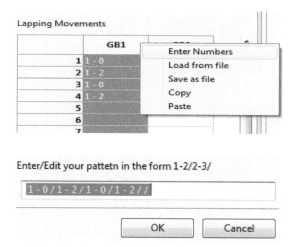

FIGURE 13.7 Lapping motion can be saved, loaded or edited in a single line.

```
Reference:
Reference source (optional)
Explanation:
Some exaplnation about this sample (optional)
```

Such files can be edited by any text editor, like Notepad, or Notepad++ etc. It is important that the keywords "Tempi:" and "Chain Notation:" are correctly written and the Tempi information is in before the chain notation.

Parts of the lapping motion can be copied and placed in another place, or repeated a number of times. For these steps, the block with the desired cells has to be selected and then the possible options appear with the right mouse button over the selection (Figure 13.8).

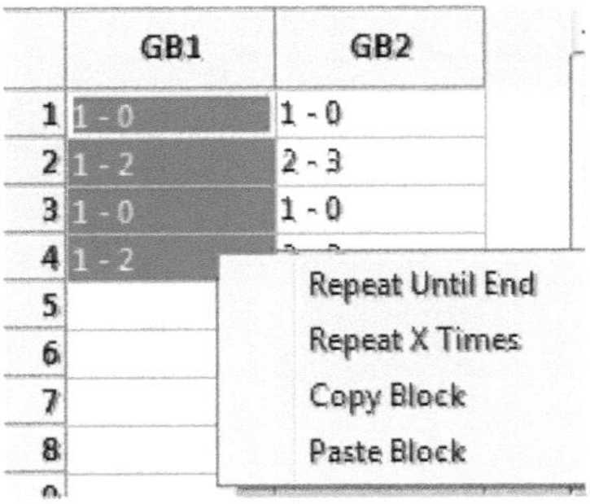

	GB1	GB2	
1	1 - 0	1 - 0	
2	1 - 2	2 - 3	
3	1 - 0	1 - 0	
4	1 - 2		Repeat Until End
5			Repeat X Times
6			Copy Block
7			Paste Block
8			

FIGURE 13.8 Additional operations over blocks of cycles.

13.6   Threading

The threading of the guide bars can be set in the following sequence (Figure 13.9):

1. First, with the corresponding cell of the yarn type, which has to be used is selected by clicking once over it with the left mouse button (Figure 13.9, pos. 1). This operation assigns this yarn type to the mouse button

2. After that, each click on the cells with the threading (Figure 13.9, pos. 2) sets the assigned colour to the corresponding guides

If a larger number of guides has to be set to one yarn, it is effective to set the first (if not empty) and the **last** cell with the mouse button and then click on the side cell (Figure 13.9, pos. A) - and all guides between the first and the last will be assigned to this colour. Alternatively, if the guide bar is selected with a tick in the editing box (Figure 13.9, pos. B), the buttons for full threading, one in - one out, can be used to set this threading for the **active area** of the guide. The active area includes the guides up to the last active one, where any yarn, or an **empty yarn** is assigned. This area is used for repeat calculations. The active area can be filled as well with more complicated threading pattern as for instance AB.BBC and repeat, with the button "R" (for repeat) and then entering the numerical representation of the threading pattern, which in this case will be 120223, where 1 represents the first yarn type, 2 the second, and 0 the empty yarn.

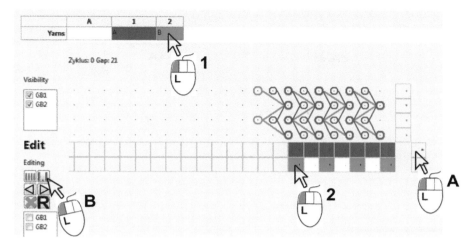

FIGURE 13.9 Setting the threading of the guide bars.

13.7 Pattern update and exports

The software updates the pattern drawing after **some** of the user actions but not after all of these. The actions, where the user can enter wrong information - for instance when a non-numerical or not knitable swinging in and swinging out positions are entered, the user input is first checked for feasibility and if not acceptable - the input is ignored or the user is asked to correct it. If the pattern is not updated in such a case, the "Redraw" button (Figure 13.10, pos. A) has to be used. The lapping diagram can be exported in various raster and vector formats with the button of the photo camera (Figure 13.10, pos. B). An Excel sheet is a useful reporting format with all pattern information (Figure 13.10, pos. C), where the pattern of the single guides, threading, take off, etc., are saved.

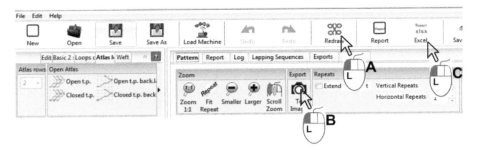

FIGURE 13.10 Setting the threading of the guide bars.

13.8   Take-off and run-in

For the computation of the run-in and for the 3D visualisation in the 3D version, the take-off values are required. These can be entered in different units as millimetres per cycle, which corresponds to the loop height for single needle bed samples and to the half loop height for the double needle bed samples. After entering any of these values, or in courses per centimetre or per inch, the tick after the text field has to be used in order for this value to be applied (Figure 13.11) with "Fill Table" button are transferred to the table. Modern machines with multi-speed drives have to be identified as such in the machine settings menu first, and then the table for the input of different values in the take-off table is activated. The software computes the yarn length per cycle per guide bar and lists this in the table, and summarizes these values as Run-in per rack (480 cycles) for each guide bar. The data for the threading of the guide bars, the machine gauge and the yarn fineness are used for calculation of the specific weight per guide bar and the total theoretical specific weight in grams per square meter for the fabrics.

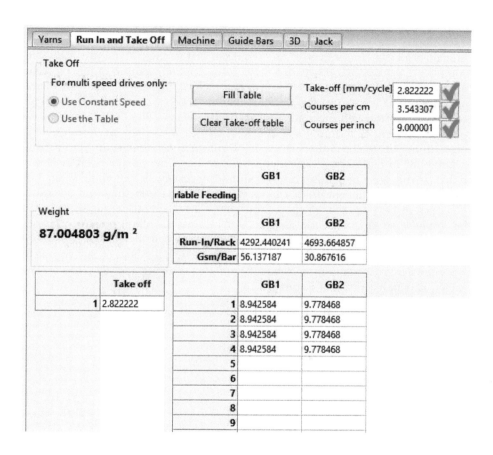

FIGURE 13.11 Take-off definition and run-in information.

13.9   3D images

The 3D visualisation is possible only in the "3D" version of the software, which is separate from the "classical 2D" version in order to allow the users, which do not change the pattern often and do not have complicated structures to obtain useful software acceptable at appropriate costs. Building a 3D representation of warp knitted fabrics is a complex modelling procedure, for which several data about the yarn properties, yarn tension during the knitting, eventual finishings of the fabrics, etc., are required. After these data are available, complex mathematical procedures have to be applied again in order to obtain a realistic structure. These procedures can include the use of Finite Elemente Method (FEM) computations, digital chain method or a combination of other contact detection and mechanics related algorithms. The TexMind software provides several functions for exports of the geometry for performing such algorithms with **external** tools like LS-Dyna, Ansys, Abaqus and has as well a few built-in algorithms, which can be used for some purposes. Which way it has to be used depends on the required accuracy of the simulation, but mainly on the qualification of the user and the available time and software. For pattern development purposes, initially, pure geometrical algorithms can be applied, so that the user can get an impression about the new fabrics **without** entering all the above explained data. Such images are generated with the button "3D" and visualised in Figure 13.12. View of the complete system with visible lapping movement numbers, drawing and 3D parts on one screen is demonstrated in Figure 13.13.

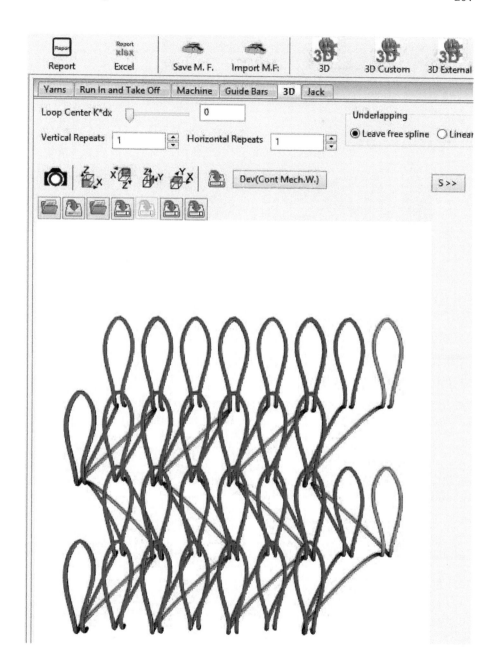

FIGURE 13.12 Panel for 3D visualisation of warp knitted structures.

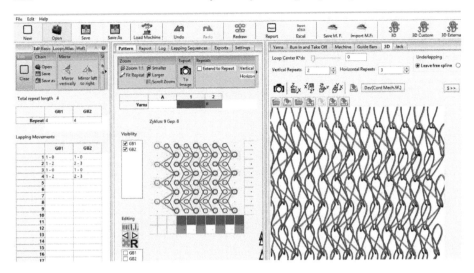

FIGURE 13.13 View of the panels with the lapping movement, drawing and 3D visualisation.

13.10   Export options of the pattern

For documentation purposes, the pattern data, the calculations and the diagrams can be exported into MS Excel sheet format. All lapping diagrams can be exported both in pixel based JPG, BMP and PNG files and vector based EMF, SVB, EPS and PDF files. For 3D geometry additionally to the high quality JPG export VRML, X3D and STL formats are integrated and allow use of the geometry into other computer graphics software for visualisation and simulations or for 3D printing.

13.11   Conclusions

The briefly described main drawing features of the software "TexMind Warp Knitting Pattern Editor 3D" demonstrated some of the functionalities used by the creation of this book and many samples for projects. The software allows a quick check of the lapping diagram, threading and helps the reader gain a first impression of the future of fabrics.

Bibliography

[1] Karl Mayer holding gmbh & co.kg.

[2] Karl Mayer Academy. *Fundamentals of Warp Kntting*. Karl Mayer Academy, Unknown.

[3] G.L. Allison. Warp knitting calculations made easy. *Silk and Rayon Record*, March 1958.

[4] Prabir Kumar Banerjee. *Principles of Fabric Formation*. CRC Press, 2018.

[5] Kazimierz Kopias; Krzysztof Kowalski; et al. *Structure and technology of warp-knitted fabrics / Budowa i technologia dzianin kolumienkowych*. Wydawnictowo Politechniki Lodzkiej, Lodz, 2010.

[6] P. Grosberg. 3—the geometry of warp-knitted fabrics. *Journal of the Textile Institute Transactions*, 51(1):T39–T48, 1960.

[7] Frank Helbig, Dietmar Reuchsel, and Frank Vettermann. Verfahren zur herstellung einer abstandswirkware sowie danach hergestellte abstandswirkware, 1995.

[8] Frank Uwe Helbig. *Druckelastische 3D-Gewirke: Gestaltungsmerkmale und mechanische Eigenschaften druckelastischer Abstandsgewirke*. Südwestdeutscher Verlag für Hochschulschriften, 2011.

[9] Rolf Hufschläger, Rohrbach Gundolf, and Stürner E. *Maschentechnik - Ketten- und Raschelwirkerei. Mustern-Fachrechnen - Musterung mittels Elektronik*. Arbeitgeberkreis Gesamttextil - Frankfurt am Main, 2 Ediction, 1986.

[10] J. Kaufmann, Y. Kyosev, H. Rabe, K. Gustke, and H. Cebulla. Investigation of the elastic properties of weft-knitted metal-reinforced narrow composites. In Anne Schwarz-Pfeiffer Yordan Kyosev, Boris Mahltig, editor, *Narrow and Smart Textiles*, pages 49–57. Springer International Publishing AG, 2018.

[11] Yordan Kyosev. Texmind warp knitting editor software, Texmind UG, 2015.

[12] Yordan Kyosev. *Topology-Based Modeling of Textile Structures and Their Joint Assemblies Principles, Algorithms and Limitations*. Springer International Publishing, 2019.

[13] Yordan K. Kyosev. *Texmind Warp Knitting Editor 3D*, 2017.

[14] Arzu Marmarali. *Gözgö Örmeciligi*. Meta Basim Matbaacihk Hizmetleri, Izmir, 2014.

[15] K. Mista. *Warp Knitting Machine with Jacquard-Control*, 1996.

[16] Kresimir Mista. Warp knitting machine with piezoelectrically controlled jacquard patterning, 1995.

[17] Peter Offermann and Harald Tausch-Marton. *Grundlagen der Maschenwarentechnologie*. Springer Vieweg, 1978.

[18] D.F. Paling. *Warp Knitting Technology*. Notthingham, 1965.

[19] S. Raz. *Warp Knitting Production*. Verlag Melliand Textilberichte GmbH, Heidelberg, 1987.

[20] W. Renkens and Y. Kyosev. Geometrical modelling of warp knitted fabrics. In *"Finite Element Modelling of Textiles and Textile Composites", St-Petersburg, 26-28 Sept. 2007, CD-ROM Proceedings*, 2007.

[21] W. Renkens and Y. Kyosev. Virtual development of warp knitted spacer fabrics for reinforced composites. In *Aachen Dresden International Textile Conference*, volume Proceedings CD-ROM, 2008.

[22] Wilfried Renkens and Yordan Kyosev. Geometry modelling of warp knitted fabrics with 3d form. *Textile Research Journal*, 81(4):437–443, 2011.

[23] Max Rogler and Martin Humboldt. *Bindungslehre der Kettenwirkerei*. UNITEX, S. Humboldt, 6051 Nieder Roden, West Germany, 1966.

[24] David Spencer. *Knitting Technology*. Nottingham, 2001.

[25] Klaus-Peter Weber. *Die Maschenbindungen der Kettenwirkerei*. Werkegemeinschaft Karl Mayer e.V., 6053 Obertshausen, 1966.

[26] Klaus Peter Weber and Marcus Weber. *Wirkeri und Strickerei*. Deutsce Fachverlag, Melliand, 2008.

[27] B. Wheatley. *Raschel Lace Production*. National Knitted Outerwear Association, 51 Madison Avenue, New York, N.Y: 10010, 1972.

[28] Chris Wilkens. *Warp Knit Fabric Construction*. U.Wilkens Verlag, Heusenstamm, Germany, 1995.

Index

Atlas
    lap, 238
atlas, 33, 54, 94
    open, 33

backlapped, 36
bar
    string, 127

Charmeuse, 85
connecting
    loop, 11
    yarn, 11
cord stitch
    partial threading, 51
cord stitch fabric
    closed, 48
    open, 48
cord-tricot, 85
counter-
    lapping, 68
Course, 17
CPC, 17
CPI, 17
crochet knitting, 23

dominance, 70
Double Atlas, 94
Double face
    fabrics, 229
Double needle bar
    needle bar, 229
Double Vandyke, 94

EAC, 182
EBC, 182
elastic
    ground, 167

electronic beam control, 182
equal
    underlapping, 132

fabric
    jacquard, 187
    single bar, 43
face
    double, 20
    single, 20
fall plate, 171
Fineness, 17

Guide bar
    numbering, 16
guide bar
    jacquard, 187

hexagon colour effect pattern, 100
hexagonal
    ground, 162

inlay, 127

jacquard, 187
jacquard fabric
    types, 188

Köpper, 37

Laid-in, 127
lap
    special, 39
lapping, 23
    1 1, 30
    equal, 68
    movement, 23
lapping width, 21

301